PERTURBATION METHODS, INSTABILITY, CATASTROPHE AND CHAOS

PERTURBATION METHODS, INSTABILITY, CATASTROPHE AND CHAOS

C F Chan Man Fong & D De Kee

Department of Chemical Engineering
Tulane University, USA

World Scientific
Singapore • New Jersey • London • Hong Kong

Published by

World Scientific Publishing Co. Pte. Ltd.

P O Box 128, Farrer Road, Singapore 912805

USA office: Suite 1B, 1060 Main Street, River Edge, NJ 07661

UK office: 57 Shelton Street, Covent Garden, London WC2H 9HE

Library of Congress Cataloging-in-Publication Data
Chan Man Fong, C.F.
    Perturbation methods, instability, catastrophe, and chaos / C.F.
Chan Man Fong, D. De Kee.
        p.   cm.
    Includes bibliographical references and indexes.
    ISBN 981023726X   ISBN 9810237278 (pbk)
    1. Differential equations -- Numerical solutions.   2. Perturbation
(Mathematics).   3. Bifurcation theory.   4. Chaotic behavior in
systems.   I. De Kee, D. (Daniel)   II. Title.
QA372.C5113   1999
515'.35--dc21   99-14412
                                                            CIP

British Library Cataloguing-in-Publication Data
A catalogue record for this book is available from the British Library.

Printed in Singapore by Uto-Print.

PREFACE

Almost all students of science and engineering have attended an introductory course on differential equations and many may be under the impression that all differential equations can be solved exactly. In reality, most of the differential equations that are encountered in practice do not have an exact solution. To solve these equations, we have to resort to various approximate methods such as asymptotic methods, numerical methods, or a combination of both. These approximate solutions can be validated by a qualitative analysis of the solutions of the differential equations.

This book is an introduction to the qualitative theory of differential equations and perturbation methods. The emphasis is on application of the theory rather than on rigorous mathematical developments. Many theorems are stated without formal proof, but an example is given to illustrate the application of the theorem. Wherever appropriate, the limitations and restrictions of the theory are stated.

We have assumed that the readers have followed a course on calculus and ordinary differential equations. New mathematical concepts, needed for an understanding of the theory, are explained in terms which are familiar to the readers.

In addition to the examples solved in the main text, there are additional examples, which are set as exercises, at the end of each chapter. The examples are taken from the physical, chemical, biological, ecological, and social sciences.

Chapter 1 considers the classical results of qualitative theory of differential equations. It is shown how important properties of the solutions of differential equations can be deduced without solving the differential equations. The conditions for the existence of periodic solutions for a certain class of differential equations are established. The stability of equilibrium points of linear differential equations in two-dimensional space is discussed. To obtain an approximate quantitative solution of a differential equation, we may try a perturbation method and this method is described in Chapter 2. It is not uncommon for the regular perturbation method to fail and several singular perturbation methods have been developed. In this chapter, both regular and singular perturbations are illustrated by solving several examples. The concept of stability introduced in Chapter 1 is extended in Chapter 3. Liapunov's indirect method is introduced and various methods of generating Liapunov functions are presented. An example of hydrodynamic stability, namely the Taylor stability problem, is considered. It is shown that at a critical Taylor number, the solution bifurcates. The topic of bifurcation is taken up in Chapter 4 and an introductory discussion on singularities and catastrophe theory is also included in this chapter. Finally, Chapter 5 considers the theory of chaos, a situation arisen out of several bifurcations.

This book covers the classical results as well as the recent developments in bifurcation and chaos. It can be used as a text for an advanced course in differential equations or an introductory course in dynamical systems. Since no advanced knowledge of mathematics is assumed, the present book can be understood by readers with only a rudimentary knowledge of calculus. The readership of this book is not confined to science and engineering students. Student of social science can also benefit from it.

This book is a development of courses taught by us at various universities and we are grateful to our former colleagues and students for their contributions. Ms. S. Boily deserves our warmest thanks for expertly typing this manuscript and Mr. J. Hinestroza for his help with the graphical work. We also very much appreciate the help of the staff at World Scientific.

C.F. Chan Man Fong
D. De Kee

New Orleans
December 1998

CONTENTS

CHAPTER 1

QUALITATIVE THEORY

1.1 INTRODUCTION

The solutions of many linear differential equations can be expressed in terms of known functions, such as exponential, Bessel, etc., whose values are tabulated. The solutions of non-linear differential equations can rarely be given in closed form. To obtain a quantitative result, it is usual to use numerical methods or perturbation methods. Numerical methods are given in Chan Man Fong et al. (1997) and perturbation methods will be discussed in the next chapter.

In this chapter, we adopt a geometric approach and this leads to a qualitative understanding of the solution of the non-linear differential equation. The foundation of the qualitative theory was laid by Poincaré at the end of the 19th century mainly for two-dimensional cases and extended to higher dimensions by Birkhoff. Currently, the qualitative theory of differential equations is an active area of research.

We recall that a differential equation of order n can be written as

$$\frac{d^n x}{dt^n} = F\left(x, \frac{dx}{dt}, \dots, \frac{d^{n-1}x}{dt^{n-1}}, t\right) \tag{1.1-1}$$

where x is the dependent variable and t is the independent variable.

Equation (1.1-1) can also be written as a set of n first order equations [Chan Man Fong et al. (1997), Chapter 7] as

$$\frac{d\underline{x}}{dt} = \underline{f}(\underline{x}, t) \tag{1.1-2a}$$

or $\quad \dot{x}_i = f_i(x_1, x_2, \dots, x_n, t) \tag{1.1-2b}$

where $\underline{x}$ is a column vector with components (x_i) and the dot denotes differentiation with respect to t.

The initial condition associated with Equation (1.1-2a) is

$$\underline{x}(t_0) = \underline{x}_0 \tag{1.1-3}$$

Equation (1.1-2a) has a unique solution in the neighborhood of $(\underline{x}_0, t_0)$ if $\underline{f}$ is continuously differentiable for some $\underline{x} \in \mathcal{D}$, $t \in \mathcal{I}$, where $\mathcal{D}$ is a domain containing $\underline{x}_0$, and $\mathcal{I}$ is an open interval containing t_0.

An example of non-uniqueness of solutions is the one-dimensional equation

$$\dot{x} = \sqrt{x} , \quad x(0) = 0 \tag{1.1-4a,b}$$

There are two solutions which satisfy Equations (1.1-4a, b). They are

$$x(t) = (t/2)^2 , \quad x(t) = 0 \tag{1.1-5a,b}$$

This non-uniqueness is due to the discontinuity of the derivative of $\sqrt{x}$ at $t = 0$.

A system where $\underline{f}$ does not depend explicitly on t is an **autonomous system**. If the solution of an autonomous system can be extended for all t in the interval $-\infty < t < \infty$, the system is a **dynamical system**. An autonomous system can be written as

$$\dot{\underline{x}} = \underline{f}(\underline{x}) \tag{1.1-6a}$$

$$\underline{x}(0) = \underline{x}_0 \tag{1.1-6b}$$

There is no loss of generality in assuming t_0 to be zero. The solution $\underline{x}(t)$ of Equations (1.1-6a, b) can be considered to be a curve in an n-dimensional space, the **phase space**, and the curve is an **integral curve**, **orbit** or **trajectory** through $\underline{x}_0$.

If $f_1(x_1, \dots, x_n) \neq 0$, the integral curves can be obtained by dividing the set of Equations (1.1-6a) by f_1 and the resulting equations are

$$\frac{dx_k}{dx_1} = \frac{f_k(x_1, \dots, x_n)}{f_1(x_1, \dots, x_n)} , \quad k = 2, \dots, n \tag{1.1-7}$$

If $f_1 = 0$ at some values of $\underline{x}$, we have to replace f_1 by f_2 or another f_i provided that $f_i \neq 0$.

If all f_i are zero at $\underline{x} = \underline{x}_e$, that is to say $\underline{f}(\underline{x}_e) = \underline{0}$, $\underline{x}_e$ is a **critical point**, an **equilibrium point**, a **singular point**, or a **stationary point**. The point $\underline{x}_e$ is an **isolated critical point** if there is no other critical point in the neighborhood of $\underline{x}_e$.

Note that at an equilibrium point, $\dot{\underline{x}} = \underline{0}$. If we associate Equations (1.1-6a, b) with a flow, the equilibrium point is a stagnation point (the velocity $\dot{\underline{x}}$ is zero).

Many processes in pure and applied sciences are autonomous systems and we list some of the properties of the solution of an autonomous system.

(i)   If $\underline{x}(t)$ is a solution of Equations (1.1-6a, b), $\underline{x}(t+a)$ is also a solution for any constant a. A trajectory represents several solutions which differ from one another by a translation of t (see Problem 1).

(ii)   Trajectories do not pass through equilibrium points.

(iii)   If a trajectory terminates at a point, that point is an equilibrium point.

(iv)   Trajectories cannot cross each other.

(v)   The trajectory of a periodic solution is a closed curve.

By examining the geometrical properties of trajectories, we can deduce certain qualitative properties, such as boundedness and periodicity, of the solutions. To facilitate the discussions, we start by considering the two-dimensional autonomous system. In this case, the phase space is a plane and the properties of the two-dimensional curves are well understood. The two-dimensional case has wide applications because many equations in mechanics and electricity are second order differential equations (see Problem 2).

1.2   TWO-DIMENSIONAL LINEAR SYSTEMS

The two-dimensional autonomous system can be written as

$$\dot{x}_1 = f_1(x_1, x_2), \qquad \dot{x}_2 = f_2(x_1, x_2) \qquad\qquad (1.2\text{-}1a,b)$$

Suppose (x_{1e}, x_{2e}) is a critical point and, by a simple shift of the origin, we can move the critical point to the origin. Expanding f_1 and f_2 about the origin as a Taylor series yields

$$f_1(x_1, x_2) = x_1\frac{\partial f_1}{\partial x_1} + x_2\frac{\partial f_1}{\partial x_2} + \frac{1}{2}\left(x_1^2\frac{\partial^2 f_1}{\partial x_1^2} + 2x_1 x_2\frac{\partial^2 f_1}{\partial x_1\,\partial x_2} + x_2^2\frac{\partial^2 f_1}{\partial x_2^2}\right) + \ldots \quad (1.2\text{-}2a)$$

$$f_2(x_1, x_2) = x_1\frac{\partial f_2}{\partial x_1} + x_2\frac{\partial f_2}{\partial x_2} + \frac{1}{2}\left(x_1^2\frac{\partial^2 f_2}{\partial x_1^2} + 2x_1 x_2\frac{\partial^2 f_2}{\partial x_1\,\partial x_2} + x_2^2\frac{\partial^2 f_2}{\partial x_2^2}\right) + \ldots \quad (1.2\text{-}2b)$$

where all the derivatives are evaluated at $(0, 0)$.

Combining Equations (1.2-1a, b, 2a, b) yields

$$\begin{bmatrix}\dot{x}_1 \\[2mm] \dot{x}_2\end{bmatrix} = \begin{bmatrix}\dfrac{\partial f_1}{\partial x_1} & \dfrac{\partial f_1}{\partial x_2} \\[3mm] \dfrac{\partial f_2}{\partial x_1} & \dfrac{\partial f_2}{\partial x_2}\end{bmatrix}\begin{bmatrix}x_1 \\[2mm] x_2\end{bmatrix} + 0\left(|\underline{x}|^2\right) \qquad\qquad (1.2\text{-}3a)$$

or $\dot{\underline{x}} = D\underline{f}(\underline{0})\underline{x} + 0(|\underline{x}|^2)$ (1.2-3b)

The elements of the Jacobian matrix $D\underline{f}(\underline{0})$ are constants and if terms of order $|\underline{x}|^2$ can be neglected, Equation (1.2-3a) simplifies to

$$\begin{bmatrix} \dot{x}_1 \\ \dot{x}_2 \end{bmatrix} = \begin{bmatrix} a_{11} & a_{12} \\ a_{21} & a_{22} \end{bmatrix} \begin{bmatrix} x_1 \\ x_2 \end{bmatrix}$$ (1.2-4a)

or $\dot{\underline{x}} = \underline{\underline{A}}\underline{x}$ (1.2-4b)

where a_{ij} are elements of $D\underline{f}(\underline{0})$ (or $\underline{\underline{A}}$).

To solve Equation (1.2-4b) we assume that the solution is of the form

$$\underline{x} = \underline{c}e^{\lambda t}$$ (1.2-5)

where $\underline{c}$ and λ are constants.

Substituting Equation (1.2-5) into Equation (1.2-4b) yields

$$(\underline{\underline{A}} - \lambda\underline{\underline{I}})\underline{c}e^{\lambda t} = \underline{0}$$ (1.2-6)

where $\underline{\underline{I}}$ is the identity matrix.

The non-trivial solution of Equation (1.2-6) implies that the determinant $|\underline{\underline{A}} - \lambda\underline{\underline{I}}| = \underline{0}$, that is to say λ is an eigenvalue of $\underline{\underline{A}}$. The two eigenvalues of $\underline{\underline{A}}$ are given by

$$\begin{vmatrix} a_{11}-\lambda & a_{12} \\ a_{21} & a_{22}-\lambda \end{vmatrix} = 0$$ (1.2-7)

The solution of Equation (1.2-7) is

$$(\lambda_1, \lambda_2) = \left[(a_{11} + a_{22}) \pm \sqrt{(a_{11} - a_{22})^2 + 4a_{12}a_{21}} \right] / 2$$ (1.2-8)

Depending on the values of a_{ij}, we have the following possibilities.

(i) $(a_{11} - a_{22})^2 + 4a_{12}a_{21} > 0$. In this case, the two roots (λ_1, λ_2) are real, distinct, and non-zero.

(ii) $(a_{11} - a_{22})^2 + 4a_{12}a_{21} < 0$. In this case, the roots are complex and, since a_{ij} are real, λ_1 and λ_2 are complex conjugates.

(iii) $(a_{11} - a_{22})^2 + 4a_{12}a_{21} = 0$. From Equation (1.2-8), we deduce that $\lambda_1 = \lambda_2$.

In case (i), the eigenvalues are distinct and the eigenvectors $\underline{v}_1$, $\underline{v}_2$ associated respectively with λ_1 and λ_2 are linearly independent.

We define a non-singular matrix $\underline{\underline{P}}$ by

$$\underline{\underline{P}} = \begin{bmatrix} \underline{v}_1 & \underline{v}_2 \end{bmatrix} \tag{1.2-9}$$

On applying the linear transformation

$$\underline{x} = \underline{\underline{P}} \, \underline{y} \tag{1.2-10}$$

Equation (1.2-4b) becomes

$$\underline{\dot{y}} = \underline{\underline{P}}^{-1} \underline{\underline{A}} \, \underline{\underline{P}} \, \underline{y} = \underline{\underline{\Lambda}} \, \underline{y} \tag{1.2-11a,b}$$

where $\underline{\underline{\Lambda}}$ is the diagonal matrix with elements λ_1 and λ_2.

The linear transformation does not change the qualitative properties of the solution and we study the solutions of Equation (1.2-11b) instead of Equation (1.2-4b).

The diagonalization of a matrix and the properties of the linear transformation are discussed in books on linear algebra [for example, Larson and Edwards (1991)].

In component form, Equation (1.2-11b) is written as

$$\dot{y}_1 - \lambda_1 y_1, \qquad \dot{y}_2 = \lambda_2 y_2 \tag{1.2-12a,b}$$

The solution is

$$y_1 = y_{10} \, e^{\lambda_1 t}, \qquad y_2 = y_{20} \, e^{\lambda_2 t} \tag{1.2-13a,b}$$

where (y_{10}, y_{20}) are the initial values of (y_1, y_2).

Eliminating t yields

$$\frac{dy_1}{dy_2} = \frac{\lambda_1 y_1}{\lambda_2 y_2} \tag{1.2-14}$$

The trajectories are obtained by solving Equation (1.2-14) and they are

$$y_1 = k\, y_2^{\lambda_1/\lambda_2} \tag{1.2-15}$$

where k is a constant and is determined by imposing the initial conditions.

If λ_1 and λ_2 are both negative, we deduce from Equations (1.2-13a, b) that both y_1 and y_2 tend to the origin as t tends to infinity. The equilibrium point $(\underline{x} = \underline{0})$ is an **asymptotically stable node**. If λ_1 and λ_2 are both positive, $|y_1|$ and $|y_2|$ tend to infinity as t tends to infinity. The equilibrium point is an **unstable node**.

In the case where λ_1 and λ_2 have different signs $(\lambda_1 > 0$ and $\lambda_2 < 0)$, $|y_1|$ tends to infinity and y_2 tends to zero as t tends to infinity. The equilibrium point is a **saddle point** and is **unstable**.

We illustrate these cases by considering numerical examples.

Example 1.2-1. **Linear systems**

Sketch the trajectories of

(a) $\dot{y}_1 = y_1, \qquad \dot{y}_2 = 2y_2$ \hfill (1.2-16a,b)

(b) $\dot{y}_1 = -2y_1, \qquad \dot{y}_2 = -y_2$ \hfill (1.2-17a,b)

(c) $\dot{y}_1 = 2y_1, \qquad \dot{y}_2 = -2y_2$ \hfill (1.2-18a,b)

The solutions of Equations (1.2-16a, b) are

$$y_1 = y_{10}\, e^t, \qquad y_2 = y_{20}\, e^{2t} \tag{1.2-19a,b}$$

As $t \longrightarrow \infty$, $(y_1, y_2) \longrightarrow (\pm\infty)$ depending on the sign of (y_{10}, y_{20}).

The trajectories are given by [Equation (1.2-15)]

$$y_1 = k\, y_2^{1/2} \tag{1.2-20a}$$

or $$y_2 = (y_1/k)^2 \tag{1.2-20b}$$

which are parabolas.

The trajectories are shown in Figure 1.2-1 and the arrows indicate the direction of increasing t. To determine k we need to impose initial conditions. If (y_{10}, y_{20}) are $(1, 1)$, k is 1 and Equation (1.2-20b) becomes

$$y_2 = y_1^2 \qquad\qquad (1.2\text{-}21)$$

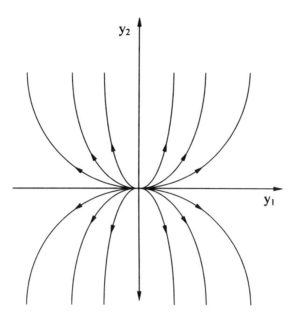

FIGURE 1.2-1   **Phase portrait of an unstable node. The arrows indicate the direction of increasing t**

Solving Equation (1.2-17a, b) yields

$$y_1 = y_{10}\, e^{-2t}, \qquad y_2 = y_{20}\, e^{-t} \qquad\qquad (1.2\text{-}22\text{a,b})$$

The functions $y_1(t)$ and $y_2(t)$ tend to the origin as t tends to infinity. The trajectories are parabolas

$$y_1^2 = k\, y_2 \qquad\qquad (1.2\text{-}23)$$

Figure 1.2-2 illustrates the trajectories.

The solution of Equations (1.2-18a, b) are

$$y_1 = y_{10}\, e^{2t}, \qquad y_2 = y_{20}\, e^{-2t} \qquad\qquad (1.2\text{-}24\text{a,b})$$

In this case, $y_1 \longrightarrow \pm\infty$ depending on the sign of y_{10} and $y_2 \longrightarrow 0$ as $t \longrightarrow \infty$. The trajectories are given by

$$y_1 = k\, y_2^{-1} \qquad\qquad (1.2\text{-}25\text{a})$$

or $$y_1 y_2 = k \qquad\qquad (1.2\text{-}25\text{b})$$

The trajectories are rectangular hyperbolas and are shown in Figure 1.2-3. Note that if y_{10} is zero, the trajectory is the y_2-axis and if y_{20} is zero, the trajectory is the y_1-axis.

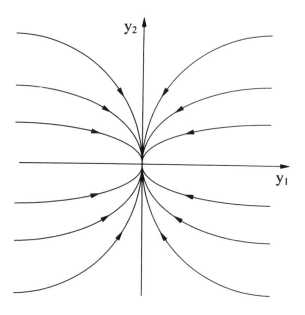

FIGURE 1.2-2   Phase portrait of a stable node. The arrows indicate the direction of increasing t

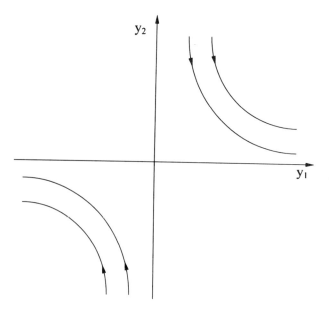

FIGURE 1.2-3   Phase portrait of a saddle point. The arrows indicate the direction of increasing t

In case (ii), λ_1 and λ_2 are complex conjugates and can be written as

$$\lambda_1 = \rho + i\omega, \qquad \lambda_2 = \rho - i\omega \qquad (1.2\text{-}26a,b)$$

Rather than working with complex numbers, we transform the matrix $\underline{A}$ in Equation (1.2-4b) to its canonical form [Barnett (1990)] and Equation (1.2-4a) becomes

$$\begin{bmatrix} \dot{y}_1 \\ \dot{y}_2 \end{bmatrix} = \begin{bmatrix} \rho & -\omega \\ \omega & \rho \end{bmatrix} \begin{bmatrix} y_1 \\ y_2 \end{bmatrix} \qquad (1.2\text{-}27)$$

It is easier to solve Equation (1.2-27) by introducing polar coordinates (r, θ) defined by

$$y_1 = r\cos\theta, \qquad y_2 = r\sin\theta \qquad (1.2\text{-}28a,b)$$

Combining Equations (1.2-27, 28a, b) yields

$$\dot{r}\cos\theta - r\dot{\theta}\sin\theta = r\,(\rho\cos\theta - \omega\sin\theta) \qquad (1.2\text{-}29a)$$

$$\dot{r}\sin\theta + r\dot{\theta}\cos\theta = r\,(\rho\sin\theta + \omega\cos\theta) \qquad (1.2\text{-}29b)$$

Multiplying Equations (1.2-29a, b) by $\cos\theta$ and $\sin\theta$ respectively and adding the resulting expressions yields

$$\dot{r} = \rho\, r \qquad (1.2\text{-}30a)$$

Similarly we can deduce that

$$\dot{\theta} = \omega \qquad (1.2\text{-}30b)$$

The solutions of Equations (1.2-30a, b) are

$$r = a\,e^{\rho t}, \qquad \theta = \omega t + c \qquad (1.2\text{-}31a,b)$$

where a and c are constants.

From Equations (1.2-30a, b), we obtain

$$\frac{dr}{d\theta} = \frac{\rho\, r}{\omega} \qquad (1.2\text{-}32)$$

The solution is

$$r = b \exp (\rho\theta / \omega) \tag{1.2-33}$$

The trajectories are spirals. If $\omega > 0$, we deduce from Equations (1.2-30b) that θ increases with time and the spirals are anti-clockwise. If $\omega < 0$, the spirals are clockwise.

Combining Equations (1.2-28a, b, 31a, b) yields

$$y_1 = a e^{\rho t} \cos (\omega t + c), \qquad y_2 = a e^{\rho t} \sin (\omega t + c) \tag{1.2-34a,b}$$

Eliminating t, we obtain the trajectories and they are given by

$$\sqrt{y_1^2 + y_2^2} = a e^{\rho t} \tag{1.2-35}$$

As shown earlier, the trajectories are spirals.

The constants a and c are determined from initial conditions. If the initial values of (y_1, y_2) are (y_{10}, y_{20}), we deduce from Equations (1.2-34a, b) that

$$a = \sqrt{y_{10}^2 + y_{20}^2}, \qquad c = \omega^{-1} \tan^{-1} (y_{20}/y_{10}) \tag{1.2-36a,b}$$

It can be seen from Equations (1.2-34a, b) that the nature of the solutions depends on ρ. If $\rho > 0$, $|y_1|$ and $|y_2|$ tend to infinity as $t \longrightarrow \infty$ and the origin is an **unstable focus**. If $\rho < 0$, y_1 and y_2 tend to zero as $t \longrightarrow \infty$ and the origin is an **asymptotic stable focus**. If $\rho = 0$, the trajectories are circles and the origin is not on the trajectories. In this case, the origin is a **stable center**.

The next example examines the case of a stable focus.

***Example 1.2-2.* Stability of equilibrium points**

Discuss the nature of the equilibrium point of the system

$$\begin{bmatrix} \dot{x}_1 \\ \dot{x}_2 \end{bmatrix} = \begin{bmatrix} 1 & -4 \\ 2 & -3 \end{bmatrix} \begin{bmatrix} x_1 \\ x_2 \end{bmatrix} \tag{1.2-37}$$

The origin is the equilibrium point of the system. The eigenvalues of the matrix are given by

$$-(1 - \lambda)(3 + \lambda) + 8 = 0 \tag{1.2-38}$$

The solutions are

$$\lambda_1 = -1 + 2i, \qquad \lambda_2 = -1 - 2i \qquad\qquad (1.2\text{-}39\text{a,b})$$

The eigenvector $\underline{v}_1$, with components (v_{11}, v_{12}), is obtained by solving

$$\begin{bmatrix} 1+1-2i & -4 \\ 2 & -3+1-2i \end{bmatrix} \begin{bmatrix} v_{11} \\ v_{12} \end{bmatrix} = \begin{bmatrix} 0 \\ 0 \end{bmatrix} \qquad\qquad (1.2\text{-}40)$$

Equation (1.2-40) can be written as two linearly dependent equations and the upper one can be written as

$$(2 - 2i) v_{11} - 4 v_{12} = 0 \qquad\qquad (1.2\text{-}41)$$

Solving Equation (1.2-41) yields

$$\underline{v}_1 = \begin{bmatrix} 2 \\ 1 - i \end{bmatrix} = \begin{bmatrix} 2 \\ 1 \end{bmatrix} + i \begin{bmatrix} 0 \\ -1 \end{bmatrix} \qquad\qquad (1.2\text{-}42\text{a,b})$$

We choose the transformation matrix $\underline{\underline{P}}$ to be

$$\underline{\underline{P}} = \begin{bmatrix} \text{Im}(\underline{v}_1) & \text{Re}(\underline{v}_1) \end{bmatrix} = \begin{bmatrix} 0 & 2 \\ -1 & 1 \end{bmatrix} \qquad\qquad (1.2\text{-}43\text{a,b})$$

where Im and Re denote respectively the imaginary and real parts.

Combining Equations (1.2-10, 11a, 43b) yields

$$\begin{bmatrix} \dot{y}_1 \\ \dot{y}_2 \end{bmatrix} = \begin{bmatrix} -1 & -2 \\ 2 & -1 \end{bmatrix} \begin{bmatrix} y_1 \\ y_2 \end{bmatrix} \qquad\qquad (1.2\text{-}44)$$

Note that Equations (1.2-27, 44) are of the same form and the solutions are

$$y_1 = a e^{-t} \cos(2t + c), \qquad y_2 = a e^{-t} \sin(2t + c) \qquad\qquad (1.2\text{-}45\text{a,b})$$

The origin is a stable focus and the trajectories are shown in Figure 1.2-4.

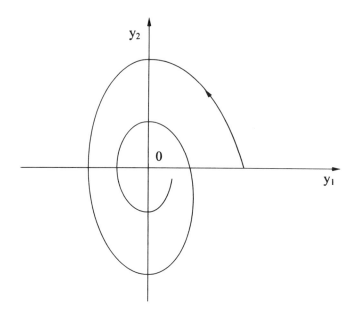

FIGURE 1.2-4   Phase portrait of a stable focus. The arrows indicate the direction of increasing t

Using Equations (1.2-10, 43b, 45a, b), we deduce that

$$x_1 = 2a e^{-t} \sin (2t + c), \qquad x_2 = a e^{-t} [\sin (2t + c) - \cos (2t + c)] \qquad \text{(1.2-46a, b)}$$

The trajectories are spirals and the equilibrium point is a stable focus. The qualitative properties of the system are invariant with respect to the linear transformation.

●

In case (iii), the two roots are equal $(\lambda_1 = \lambda_2 = \lambda)$ and we are not guaranteed that we can have two linearly independent eigenvectors. Depending on the availability of independent eigenvectors, the two possible canonical forms in this case are [Barnett (1990)]

$$\begin{bmatrix} \lambda & 0 \\ 0 & \lambda \end{bmatrix} \quad \text{or} \quad \begin{bmatrix} \lambda & 1 \\ 0 & \lambda \end{bmatrix}$$

The equations we need to examine are

$$\dot{y}_1 = \lambda y_1, \qquad \dot{y}_2 = \lambda y_2 \qquad \qquad \text{(1.2-47a, b)}$$

$$\text{or} \qquad \dot{y}_1 = \lambda y_1 + y_2, \qquad \dot{y}_2 = \lambda y_2 \qquad \qquad \text{(1.2-48a,b)}$$

The solutions of Equations (1.2-47a, b) are

$$y_1 = y_{10} e^{\lambda t}, \qquad y_2 = y_{20} e^{\lambda t} \qquad\qquad (1.2\text{-}49a,b)$$

The trajectories are straight lines and are given by

$$y_1 = (y_{10}/y_{20}) y_2 \qquad\qquad (1.2\text{-}50)$$

If $\lambda < 0$, the trajectories tend to the origin as $t \longrightarrow \infty$ and the origin is an **asymptotic stable node**. If $\lambda > 0$, the trajectories tend to infinity as $t \longrightarrow \infty$ and the origin is an **unstable node**. Some authors refer to these nodes as **stars**.

The solutions of Equations (1.2-48a, b) are

$$y_1 = y_{10} e^{\lambda t} + t y_{20} e^{\lambda t}, \qquad y_2 = y_{20} e^{\lambda t} \qquad\qquad (1.2\text{-}51a,b)$$

The equation for the trajectories is

$$\frac{dy_1}{dy_2} = \frac{\lambda y_1 + y_2}{\lambda y_2} \qquad\qquad (1.2\text{-}52)$$

The solution of this homogeneous equation is

$$\lambda y_1 = y_2 \, \ell n \, y_2 + c \, y_2 \qquad\qquad (1.2\text{-}53)$$

where c is a constant $[= (\lambda y_{10} - y_{20} \ell n \, y_{20})/y_{20}]$.

From Equations (1.2-51a, b, 52), we deduce that as $t \longrightarrow \infty$, $(dy_1/dy_2) \longrightarrow \infty$ and this implies that the trajectories are asymptotic to the y_1-axis. If $y_{20} = 0$, the trajectory is the y_1-axis.

If $\lambda < 0$, y_1 and y_2 tend to the origin as $t \longrightarrow \infty$ and the origin is an **asymptotic stable node**. If $\lambda > 0$, the origin is an **unstable node**. Some authors classify these nodes as **improper nodes**.

In the next example, we show how the canonical form of the matrix can be obtained in the case of a double root.

Example 1.2-3. **Canonical form of a matrix with double roots**

Obtain the solution of the system

$$\begin{bmatrix} \dot{x}_1 \\ \dot{x}_2 \end{bmatrix} = \begin{bmatrix} 11 & 25 \\ -4 & -9 \end{bmatrix} \begin{bmatrix} x_1 \\ x_2 \end{bmatrix} \qquad\qquad (1.2\text{-}54)$$

The eigenvalues of the matrix are given by

$$- (11 - \lambda)(9 + \lambda) + 100 = 0 \tag{1.2-55}$$

The solution is

$$\lambda = 1 \text{ (double)} \tag{1.2-56}$$

In this case, we have only one eigenvector $\underline{v}_1$ which is proportional to $[5, -2]^{\dagger}$, where $\dagger$ denotes the transpose. The matrix cannot be diagonalized. To ensure that the upper off diagonal element in the canonical form of the matrix is unity, we proceed as follows. From the Cayley-Hamilton theorem we deduce that for any (2×2) matrix $\underline{\underline{A}}$ with double root λ and for any vector $\underline{v}$, we have

$$\left[\underline{\underline{A}} - \lambda \underline{\underline{I}} \right]^2 \underline{v} = \underline{0} \tag{1.2-57}$$

We define $\underline{v}_1$ by

$$\underline{v}_1 = \left[\underline{\underline{A}} - \lambda \underline{\underline{I}} \right] \underline{v} \tag{1.2-58}$$

From Equations (1.2-57, 58) we note that $\underline{v}_1$ is an eigenvector. The transformation matrix $\underline{\underline{P}}$ is chosen to be $[\underline{v}_1, \underline{v}]$. We have assumed that $\underline{v}$ and $\underline{v}_1$ are linearly independent and $\underline{v}_1$ is not a null vector.

In our example, we choose $\underline{v}$ to be $[1, 0]^{\dagger}$ and $\underline{v}_1$ is given by

$$\underline{v}_1 = \begin{bmatrix} 10 & 25 \\ -4 & -10 \end{bmatrix} \begin{bmatrix} 1 \\ 0 \end{bmatrix} = \begin{bmatrix} 10 \\ -4 \end{bmatrix} \tag{1.2-59a,b}$$

The matrix $\underline{\underline{P}}$ is

$$\underline{\underline{P}} = \begin{bmatrix} 10 & 1 \\ -4 & 0 \end{bmatrix} \tag{1.2-60}$$

Combining Equations (1.2-10, 11a, 54) yields

$$\begin{bmatrix} \dot{y}_1 \\ \dot{y}_2 \end{bmatrix} = \begin{bmatrix} 1 & 1 \\ 0 & 1 \end{bmatrix} \begin{bmatrix} y_1 \\ y_2 \end{bmatrix} \tag{1.2-61}$$

The solutions of Equation (1.2-61) are

$$y_1 = (y_{10} + t\, y_{20})\, e^t, \qquad y_2 = y_{20}\, e^t \tag{1.2-62a,b}$$

The trajectories are given by

$$y_1 = y_2\, \ell n\, y_2 + c\, y_2 \tag{1.2-63}$$

From Equations (1.2-52, 62a, b), we deduce the following properties.

(i)   As $t \longrightarrow \infty$, both $|y_1|$ and $|y_2|$ tend to infinity and the ratio (y_1/y_2) tends to infinity. The origin is an unstable node and the trajectories are asymptotic to the y_1-axis.

(ii)   The slope (dy_1/dy_2) vanishes along the line $y_1 = -y_2$.

Figure 1.2-5 illustrates the trajectories.

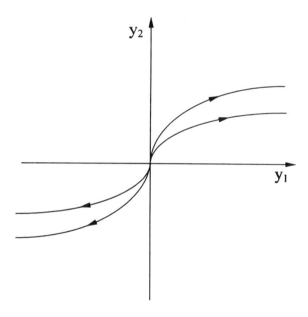

FIGURE 1.2-5   Phase portrait of an unstable improper node. The arrows indicate the direction of increasing t

Equation (1.2-8) can also be written in terms of $\text{tr}\,\underline{\underline{A}}$ (trace of $\underline{\underline{A}}$) and $\det\underline{\underline{A}}$ (determinant of $\underline{\underline{A}}$) as

$$(\lambda_1, \lambda_2) = \left[\text{tr}\,\underline{\underline{A}} \pm \sqrt{(\text{tr}\,\underline{\underline{A}})^2 - 4\det\underline{\underline{A}}} \right] / 2 \tag{1.2-64}$$

The results we have deduced for the linear system can be summarized as follows.

(i) $(\text{tr } \underline{A})^2 > 4 \det \underline{A} > 0$, the eigenvalues are real. If $\text{tr } \underline{A} < 0$, the equilibrium point is an asymptotic stable node. If $\text{tr } \underline{A} > 0$, the equilibrium point is an unstable node.

(ii) $4 \det \underline{A} > (\text{tr } \underline{A})^2 \geq 0$, the eigenvalues are complex. If $\text{tr } \underline{A} < 0$, the origin is an asymptotic stable focus. If $\text{tr } \underline{A} > 0$, the origin is an unstable focus. If $\text{tr } \underline{A} = 0$, the origin is a stable center.

(iii) $(\text{tr } \underline{A})^2 - 4 \det \underline{A} = 0$, the eigenvalues are equal. The origin is asymptotically stable if $\text{tr } \underline{A} < 0$ and unstable if $\text{tr } \underline{A} > 0$.

In all three cases, $\det \underline{A} > 0$. If $\det \underline{A} < 0$, the eigenvalues are real but have different signs and the origin is a saddle point.

We have defined the equilibrium point to be asymptotically stable if the trajectories tend to the equilibrium point as t tends to infinity. If the trajectories tend to infinity as t tends to infinity, the equilibrium point is unstable. If the trajectories are bounded, the equilibrium point (center) is stable. In Chapter 3, the definitions and concepts of stability will be amplified.

If the real parts of the eigenvalues of $D \underline{f}(\underline{0})$ [Equation (1.2-3b)] are non-zero, the system is **hyperbolic**.

By linearizing f_1 and f_2 in Equations (1.2-1a, b), we have been able to obtain explicit solutions for the system. In the next section, we extend these results to non-linear systems. For these systems, it is generally impossible to obtain exact solutions.

1.3   TWO-DIMENSIONAL ALMOST LINEAR SYSTEMS

We assume that the origin is an equilibrium point of the system defined by Equations (1.2-1a, b) and we are interested in the behavior of the trajectories of the system in the neighborhood of the origin. We further assume that f_1 and f_2 and their first partial derivatives are continuous near the origin and Equations (1.2-1a, b) can be written as [see Equations (1.2-2a) to (1.2-4b)]

$$\dot{\underline{x}} = \underline{A}\,\underline{x} + \underline{h}(\underline{x}) \tag{1.3-1}$$

The matrix $\underline{A}$ is non-singular and $\underline{h}(\underline{x})$ satisfies the condition

$$\|\underline{h}(\underline{x})\| / \|\underline{x}\| \longrightarrow 0 \quad \text{as} \quad \underline{x} \longrightarrow 0 \tag{1.3-2a}$$

or $$\lim_{r \to 0} h_i(x_1, x_2)/r = 0, \quad i = 1, 2 \tag{1.3-2b}$$

where $\| \underline{x} \| = r = \sqrt{x_1^2 + x_2^2}$ (1.3-2c,d)

The conditions we have imposed implied that the origin is an isolated equilibrium point and $\| \underline{h}(\underline{x}) \|$ is small compared to $\| \underline{x} \|$. Equations (1.3-1) to (1.3-2d) describe an **almost linear system**.

We can expect that in most cases the trajectories near the equilibrium point for the linear and almost linear systems are similar. The two possible exceptions occur when the eigenvalues of $\underline{A}$ are purely imaginary or are equal. A small change in the system will contribute a real part to the eigenvalues and the center is transformed to a stable or an unstable spiral. Likewise, under the influence of a small perturbation, the two equal real eigenvalues may become two unequal real or complex eigenvalues. We compare the type and the stability of the equilibrium of the linear and almost linear systems in Table 1.3-1. The eigenvalues of $\underline{A}$ are denoted by λ_1 and λ_2.

TABLE 1.3.1

Type and stability of equilibrium points of linear and almost linear systems

	LINEAR		ALMOST LINEAR	
	Type	Stability	Type	Stability
$\lambda_1 < \lambda_2 < 0$	node	asymptotically stable	node	asymptotically stable
$\lambda_1 > \lambda_2 > 0$	node	unstable	node	unstable
$\lambda_2 < 0 < \lambda_1$	saddle	unstable	saddle	unstable
$\lambda_{1,2} = \rho \pm i\omega$				
$\rho < 0$	spiral	asymptotically stable	spiral	asymptotically stable
$\rho > 0$	spiral	unstable	spiral	unstable
$\lambda_{1,2} = \pm i\omega$	center	stable	center or spiral	indeterminate
$\lambda_1 = \lambda_2 > 0$	node	unstable	node or spiral	unstable
$\lambda_1 = \lambda_2 < 0$	node	asymptotically stable	node or spiral	asymptotically stable

The proof of the results given in Table 1.3-1 can be found in Cesari (1971). Instead of reproducing the proof here, we shall consider a few examples.

Example 1.3-1. **Almost linear systems**

Find the equilibrium points of the system

$$\dot{x}_1 = x_2 - \mu x_1 (x_1^2 + x_2^2) \tag{1.3-3a}$$

$$\dot{x}_2 = -x_1 - \mu x_2 (x_1^2 + x_2^2) \tag{1.3-3b}$$

and determine their type and stability.

The equilibrium points are given by

$$x_2 - \mu x_1 (x_1^2 + x_2^2) = -x_1 - \mu x_2 (x_1^2 + x_2^2) = 0 \tag{1.3-4a, b}$$

The only solution is

$$x_1 = x_2 = 0 \tag{1.3-5a,b}$$

The origin is the only equilibrium point of this almost linear system.

Linearizing Equations (1.3-3a, b) yields

$$\begin{bmatrix} \dot{x}_1 \\ \dot{x}_2 \end{bmatrix} = \begin{bmatrix} 0 & 1 \\ -1 & 0 \end{bmatrix} \begin{bmatrix} x_1 \\ x_2 \end{bmatrix} \tag{1.3-6}$$

The eigenvalues are

$$\lambda_{1,2} = \pm i \tag{1.3-7}$$

The equilibrium point of the linear system is a stable center. To solve the almost linear system we introduce polar coordinates [Equations (1.2-28a, b)]. Equations (1.3-3a, b) become

$$\dot{r} \cos\theta - r\dot{\theta} \sin\theta = r \sin\theta - \mu r^3 \cos\theta \tag{1.3-8a}$$

$$\dot{r} \sin\theta + r\dot{\theta} \cos\theta = -r \cos\theta - \mu r^3 \sin\theta \tag{1.3-8b}$$

Multiplying Equations (1.3-8a, b) respectively by $\cos\theta$ and $\sin\theta$ and adding the resulting expressions yields

$$\dot{r} = -\mu r^3 \tag{1.3-9a}$$

Similarly by multiplying Equations (1.3-8a, b) respectively by $(-\sin\theta)$ and $\cos\theta$ and adding the resulting expressions, we obtain

$$\dot{\theta} = -1 \tag{1.3-9b}$$

The solutions of Equations (1.3-9a, b) are

$$r = r_0 / \sqrt{1 + 2\mu t\, r_0^2} \tag{1.3-10a}$$

$$\theta = \theta_0 - t \tag{1.3-10b}$$

where r_0 and θ_0 are respectively the initial values of r and θ.

In the almost linear system, the trajectories are clockwise spirals. If $\mu > 0$, for all $t > 0$, r decreases with increasing t and as $t \longrightarrow \infty$, $r \longrightarrow 0$. The origin is a stable focus. If $\mu < 0$, for all $t > 0$, r increases with increasing t and as $t \longrightarrow 1/2\,|\mu|\,r_0^2$, $r \longrightarrow \infty$. The origin is an unstable equilibrium point. If $\mu = 0$, the system is linear, the trajectories are circles ($r = r_0$). The center of the linear system becomes a stable spiral if $\mu > 0$ and an unstable spiral if $\mu < 0$ in the almost linear system.

It should be emphasized that the results given in Table 1.3-1 are valid only for trajectories near the equilibrium point. If the equilibrium point is an unstable focus, the trajectories for both linear and almost linear systems spiral away from the origin. For the linear system, the trajectories tend to infinity as $t \longrightarrow \infty$. They may tend to a closed curve (a periodic solution) for the almost linear system. This closed curve is a **limit cycle** and is illustrated in the next example.

Example 1.3-2. **Stability of an almost linear system**

Determine the type and stability of the equilibrium point of the system

$$\dot{x}_1 = x_2 + x_1 - \mu x_1 (x_1^2 + x_2^2) \tag{1.3-11a}$$

$$\dot{x}_2 = -x_1 + x_2 - \mu x_2 (x_1^2 + x_2^2), \quad \mu \geq 0 \tag{1.3-11b}$$

The origin is the only equilibrium point for the linear system which is

$$\begin{bmatrix} \dot{x}_1 \\ \dot{x}_2 \end{bmatrix} = \begin{bmatrix} 1 & 1 \\ -1 & 1 \end{bmatrix} \begin{bmatrix} x_1 \\ x_2 \end{bmatrix}$$
(1.3-12)

The eigenvalues are

$$\lambda_{1,2} = 1 \pm i$$
(1.3-13)

The equilibrium point is an unstable spiral. That is, as $t \longrightarrow \infty$, the trajectories tend to infinity.

To solve Equations (1.3-11a, b) we introduce polar coordinates and proceed as in Example 1.3-1. The resulting equations for r and θ are

$$\dot{r} = r(1 - \mu r^2)$$
(1.3-14a)

$$\dot{\theta} = -1$$
(1.3-14b)

Equation (1.3-14a) can be written as

$$\int \left[\frac{1}{r} + \frac{\sqrt{\mu}}{2(1 - r\sqrt{\mu})} - \frac{\sqrt{\mu}}{2(1 - r\sqrt{\mu})} \right] dr = \int dt$$
(1.3-14c)

Integrating Equations (1.3-14b, c) yields

$$\theta = \theta_0 - t$$
(1.3-15a)

$$r = r_0 / \sqrt{\mu r_0^2 + (1 - \mu r_0^2) e^{-2t}}$$
(1.3-15b)

where θ_0 and r_0 are the initial values of θ and r.

For the almost linear system, the equilibrium point is a clockwise spiral with r increasing with increasing t. From Equation (1.3-15b), we deduce that as $t \longrightarrow \infty$, $r \longrightarrow 1/\sqrt{\mu}$. If $\mu = 0$, we recover the linear system and $r \longrightarrow \infty$ as $t \longrightarrow \infty$.

We note from Equation (1.3.14a) that the two points at which $\dot{r} = 0$ are $r = 0$ and $r = 1/\sqrt{\mu}$. Thus, a trajectory that starts near the origin spirals out to the circle of radius $1/\sqrt{\mu}$ as shown in Figure 1.3-1. We now examine the stability of this limit cycle (periodic solution). If $r_0^2 > 1/\mu$, we deduce from Equation (1.3-15b) [or Equation (1.3-14a)] that r decreases as t increases and as $t \longrightarrow \infty$, $r \longrightarrow 1/\sqrt{\mu}$. Thus, all trajectories that originate near the limit cycle ($r = 1/\sqrt{\mu}$) spiral to the limit cycle, as shown in Figure 1.3-1. The limit cycle is stable.

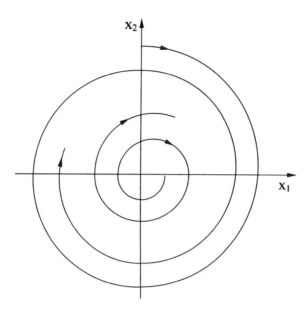

FIGURE 1.3.1   Limit cycle. The arrows indicate the direction
of increasing t

●

This example illustrates the importance of non-linear terms in modeling. The linear model predicts an unbounded solution, the almost linear model predicts a bounded solution or a periodic solution. In the next few examples, we shall compare the response of the linear and almost linear models proposed for some real systems.

Example 1.3-3. **Damped pendulum**

Investigate the motion of a damped pendulum.

The equation of motion of a damped pendulum can be written as

$$m \ell^2 \ddot{\theta} + k \ell \dot{\theta} + m g \ell \sin \theta = 0 \qquad (1.3\text{-}16)$$

where m, ℓ are the mass and the length of the pendulum respectively, θ is the angle between the pendulum and the downward vertical, g is the gravitational acceleration, and k is a positive constant.

Note that in Equation (1.3-16) we have assumed that the resistance is proportional to the velocity. In dimensionless form, Equation (1.3-16) becomes

$$\ddot{\theta} + \alpha \dot{\theta} + \omega^2 \sin \theta = 0 \qquad\qquad (1.3\text{-}17)$$

where $\alpha = k/m\ell$, $\omega^2 = g/\ell$.

Equation (1.3-17) can be written as a system of two first order equations as follows

$$x_1 = \theta \qquad\qquad (1.3\text{-}18\text{a})$$

$$\dot{x}_1 = x_2 \qquad\qquad (1.3\text{-}18\text{b})$$

$$\dot{x}_2 = -\alpha x_2 - \omega^2 \sin x_1 \qquad\qquad (1.3\text{-}18\text{c})$$

The equilibrium points of the system are given by

$$x_2 = \alpha x_2 + \omega^2 \sin x_1 = 0 \qquad\qquad (1.3\text{-}19\text{a,b})$$

The solutions are

$$x_2 = 0, \qquad\qquad x_1 = n\pi \quad (n = 0, \pm 1, \pm 2, ...) \qquad\qquad (1.3\text{-}20\text{a,b})$$

We examine the stability of the origin. Linearizing Equations (1.3-18b, c) yields

$$\begin{bmatrix} \dot{x}_1 \\ \dot{x}_2 \end{bmatrix} = \begin{bmatrix} 0 & 1 \\ -\omega^2 & -\alpha \end{bmatrix} \begin{bmatrix} x_1 \\ x_2 \end{bmatrix} \qquad\qquad (1.3\text{-}21)$$

The eigenvalues are

$$\lambda_{1,2} = \left[-\alpha \pm \sqrt{\alpha^2 - 4\omega^2} \right] / 2 \qquad\qquad (1.3\text{-}22)$$

If $\alpha^2 > 4\omega^2$ (the damping is large), $\lambda_{1,2}$ are real, negative, and distinct; the origin is an asymptotic stable node. If $\alpha^2 = 4\omega^2$, $\lambda_1 = \lambda_2 < 0$ and the origin is again an asymptotic stable node. If $\alpha^2 < 4\omega^2$, λ_1 and λ_2 are complex with negative real parts; the origin is an asymptotic stable focus.

The case $x_1 = 0$ corresponds to the pendulum being vertically downwards and, on physical grounds, we expect the origin to be stable as shown by examining the linear system.

Next, we consider the equilibrium $(\pi, 0)$. To shift $(\pi, 0)$ to $(0, 0)$ we write

$$x_1^* = x_1 - \pi, \qquad\qquad x_2^* = x_2 \qquad\qquad (1.3\text{-}23\text{a,b})$$

Substituting Equations (1.3-23a, b) into Equations (1.3-18b, c) yields

$$\dot{x}_1^* = x_2^* \tag{1.3-24a}$$

$$\dot{x}_2^* = -\alpha x_2^* - \omega^2 \sin(x_1^* + \pi) = -\alpha x_2^* + \omega^2 \sin x_1^* \tag{1.3-24b,c}$$

The linear form of Equations (1.3-24a, c) is

$$\begin{bmatrix} \dot{x}_1^* \\ \dot{x}_2^* \end{bmatrix} = \begin{bmatrix} 0 & 1 \\ \omega^2 & -\alpha \end{bmatrix} \begin{bmatrix} x_1 \\ x_2 \end{bmatrix} \tag{1.3-24d}$$

The eigenvalues are

$$\lambda_{1,2} = \left[-\alpha \pm \sqrt{\alpha^2 + 4\omega^2} \right] / 2 \tag{1.3-25}$$

In this case, λ_1 and λ_2 are real and are of opposite sign. The equilibrium point is a saddle point and is unstable regardless of the magnitude of the damping. The present equilibrium point corresponds to a vertical upwards pendulum and, from physical consideration, we expect this position to be unstable.

Since $\sin x_1$ is periodic of period 2π, we deduce that the equilibrium points $(\pm 2n\pi, 0)$ are asymptotically stable and $[\pm (2n + 1)\pi, 0]$ are unstable. In the absence of damping $(\alpha = 0)$, Equations (1.3-22, 25) show that the equilibrium points $(2n\pi, 0)$ are centers (purely imaginary eigenvalues) and $[(2n + 1)\pi, 0]$ are saddle points (see Problem 4).

The trajectories in the phase plane for the linear system are shown in Figure 1.3-2. To obtain the trajectories for the almost linear system, we solve Equations (1.3-18b, c). Eliminating t yields

$$\frac{dx_1}{dx_2} = -\left[\frac{x_2}{\alpha x_2 + \omega^2 \sin x_1} \right] \tag{1.3-26}$$

Equation (1.3-26) is not easy to solve. However, From Table 1.3-1, we deduce that for the almost linear system the origin $(0, 0)$ is an asymptotic stable node, $(-\pi, 0)$ and $(\pi, 0)$ are saddle points. All trajectories near $(0, 0)$ tend to $(0, 0)$ as $t \longrightarrow \infty$ and there is only one trajectory that passes through $(\pi, 0)$. This trajectory is a **separatrix**. Similarly, there is another separatrix that passes through $(-\pi, 0)$. All trajectories that originate in the region bounded by these two separatrices are attracted to the origin and this region is the **basin of attraction** of the origin. Similar arguments can be applied to the region $\pi \le x \le 3\pi$, $-3\pi \le x \le -\pi$, and the phase portrait for the almost linear system is shown in Figure 1.3-3.

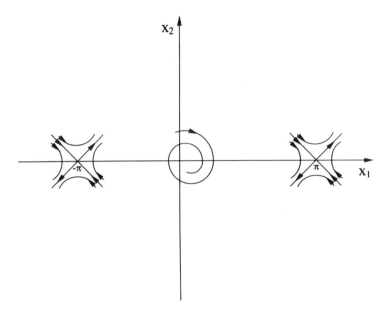

FIGURE 1.3-2   Phase portrait of a linear damped pendulum

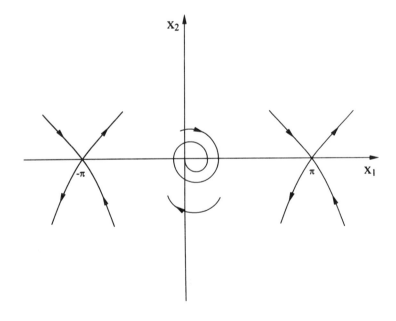

FIGURE 1.3.3   Phase portrait of a non-linear damped oscillator

Example 1.3-4. **Lotka-Volterra equation**

Discuss the qualitative solution of the Lotka-Volterra equation.

To explain the oscillatory level of certain fish caught in the Adriatic Sea, Volterra proposed the **predator-prey** model [Equations (1.3-27a, b)]. Let N_1 and N_2 be the population of the prey and the predator respectively. In the absence of predator, the rate of increase of N_1 is proportional to N_1 and in the absence of prey, the predator dies at a rate proportional to N_2. The number of encounters between the predator and the prey is proportional to $N_1 N_2$. At each such encounter, the predator devours the prey resulting in a decrease of N_1 and in an increase in N_2. Based on these assumptions, we deduce that

$$\frac{dN_1}{dt} = a_1 N_1 - b_1 N_1 N_2 \qquad (1.3\text{-}27a)$$

$$\frac{dN_2}{dt} = b_2 N_1 N_2 - a_2 N_2 \qquad (1.3\text{-}27b)$$

where a_1, a_2, b_1, and b_2 are positive constants.

Equations (1.3-27a, b) were proposed independently by Lotka and a discussion on these equations is given in Murray (1993) and Lotka (1956).

Murray (1993) non-dimensionalized Equations (1.3-27a, b) by introducing the following quantities

$$x_1 = \frac{N_1 b_2}{a_2}, \qquad x_2 = \frac{N_2 b_1}{a_1}, \qquad \tau = t a_1 \qquad (1.3\text{-}28a,b,c)$$

Substituting Equations (1.3-28a, b, c) into Equations (1.3-27a, b) yields

$$\frac{dx_1}{d\tau} = x_1 (1 - x_2) \qquad (1.3\text{-}29a)$$

$$\frac{dx_2}{d\tau} = \alpha x_2 (x_1 - 1) \qquad (1.3\text{-}29b)$$

where $\alpha = a_2/a_1$.

The equilibrium points are $(0, 0)$ and $(1, 1)$. The linearized system near $(0, 0)$ is

$$\begin{bmatrix} \dfrac{dx_1}{d\tau} \\[2ex] \dfrac{dx_2}{d\tau} \end{bmatrix} = \begin{bmatrix} 1 & 0 \\ 0 & -\alpha \end{bmatrix} \begin{bmatrix} x_1 \\ x_2 \end{bmatrix} \qquad (1.3\text{-}30)$$

The eigenvalues are

$$\lambda_1 = 1, \qquad \lambda_2 = -\alpha \tag{1.3-31a,b}$$

The origin is a saddle point and is unstable. Equation (1.3-30) shows that x_1 grows exponentially with time and x_2 decays exponentially.

To examine the solution near $(1, 1)$, we shift the origin and write

$$x_1^* = x_1 - 1, \qquad x_2^* = x_2 - 1 \tag{1.3-32a,b}$$

Equations (1.3-29a, b) become

$$\frac{dx_1^*}{d\tau} = -x_2^* (1 + x_1^*) \tag{1.3-33a}$$

$$\frac{dx_2^*}{d\tau} = \alpha x_1^* (1 + x_2^*) \tag{1.3-33b}$$

Linearizing yields

$$\begin{bmatrix} \dfrac{dx_1^*}{d\tau} \\[4mm] \dfrac{dx_2^*}{d\tau} \end{bmatrix} = \begin{bmatrix} 0 & -1 \\[4mm] \alpha & 0 \end{bmatrix} \begin{bmatrix} x_1^* \\[4mm] x_2^* \end{bmatrix} \tag{1.3-34}$$

The eigenvalues are

$$\lambda_{1,2} = \pm i \sqrt{\alpha} \tag{1.3-35}$$

The equilibrium point $(1, 1)$ is a center, it is stable and the solution is periodic. From Equation (1.3-34) we deduce that near $(1, 1)$ the trajectories are ellipses and are given by

$$\alpha (x_1^*)^2 + (x_2^*)^2 = \text{constant} \tag{1.3-36}$$

From Table (1.3-1) we note that for the almost linear system [Equations (1.3-29a, b)], the origin $(0, 0)$ is unstable and no definite conclusion can be inferred for the equilibrium point $(1, 1)$. The trajectories are given by

$$\frac{dx_1}{dx_2} = \frac{x_1 (1 - x_2)}{\alpha x_2 (x_1 - 1)} \tag{1.3-37}$$

The solution is

$$\alpha \, (x_1 - \ln x_1) \; = \; \ln x_2 - x_2 + \ln C \tag{1.3-38a}$$

or $\quad C \, x_1^{\alpha} \, x_2 \; = \; e^{\alpha x_1} \, e^{x_2}$ $\hfill$ (1.3-38b)

where C is a constant depending on the initial conditions.

The plots of Equation (1.3-38b) for fixed values of C are closed curves surrounding the equilibrium point $(1, 1)$ as shown in Figure 1.3-4. In this example, both the linear and the almost linear systems predict periodic solutions. A slight change in the initial conditions leads to a different trajectory (solution) and this is in contrast to the limit cycle discussed in Example 1.3-2. This system is structurally unstable and this concept is explained further in Chapter 3.

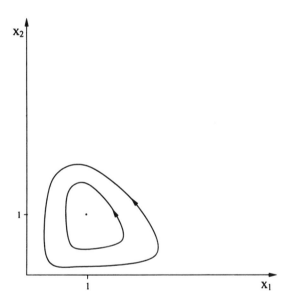

FIGURE 1.3-4   Trajectories of the Lotka-Volterra equation

Various authors have used the Lotka-Volterra model to interpret the periodic variations of the population of Canadian lynx and snow-shoe hare during the period 1845–1935. A critical assessment of the suitability of the model is given in Murray (1993).

One criticism of the present model is that in the absence of the predator, the population of the prey increases exponentially to infinity. This Malthusian growth is not observed in practice. The **logistic equation** proposed by Verhulst conforms better to the reality and is written as

$$\frac{dN_1}{dt} \; = \; a_1 N_1 \left(1 - \frac{N_1}{K} \right) \tag{1.3-39}$$

where K is a positive constant.

The two equilibrium points of Equation (1.3-39) are

$$N_1 = 0 , \qquad N_1 = K \tag{1.3-40a,b}$$

The equilibrium point $(N_1 = 0)$ is unstable. To investigate the nature of the solution near the second equilibrium point, we write

$$N_1^* = N_1 - K \tag{1.3-41}$$

Equation (1.3-39) becomes

$$\frac{dN_1^*}{dt} = -\frac{a_1 N_1^*}{K} (N_1^* + K) \approx -a_1 N_1^* \tag{1.3-42a,b}$$

The function N_1^* decays exponentially to zero and the equilibrium point $(N_1 = K)$ is asymptotically stable. If the initial population is $N_{10} < K$, N_1 increases with increasing t until $N_1 = K$ and thereafter, it remains constant. If $N_{10} > K$, N_1 is a decreasing function of t and tends to the finite limit K. The final population is K irrespective of the initial size of the population and K is the **carrying capacity** of the environment.

Equation (1.3-39) can be solved exactly and the solution is

$$\ell n\, N_1 - \ell n\, (K - N_1) = a_1 t + \ell n\, N_{10} - \ell n\, (K - N_{10}) \tag{1.3-43a}$$

or $\quad N_1 = K N_{10} / [N_{10} + (K - N_{10})\, e^{-a_1 t}] \tag{1.3-43b}$

From Equation (1.3-43b), we can verify the conclusions drawn from the linear system.

Replacing the exponential growth term in Equation (1.3-27a) by the logistic equation yields

$$\frac{dN_1}{dt} = a_1 N_1 \left(1 - \frac{N_1}{K}\right) - b_1 N_1 N_2 \tag{1.3-44}$$

Combining Equations (1.3-28a–c, 44) yields

$$\frac{dx_1}{d\tau} = x_1 (1 - \beta x_1 - x_2) \tag{1.3-45}$$

$$\frac{dx_2}{d\tau} = \alpha x_2 (x_1 - 1) \tag{1.3-29b}$$

where $\beta = a_2 / K b_2$.

For convenience, we have reproduced Equation (1.3-29b) here. The equilibrium points are $(0, 0)$, $(1/\beta, 0)$, and $(1, 1-\beta)$. The origin has been shown to be a saddle point and is unstable. To examine the nature of $(1/\beta, 0)$, we translate the origin to $(1/\beta, 0)$ by writing

$$\bar{x}_1 = x_1 - 1/\beta, \qquad \bar{x}_2 = x_2 \qquad\qquad\qquad (1.3\text{-}46a,b)$$

Substituting Equations (1.3-46a, b) into Equations (1.3-45, 29b) and retaining only the linear terms yields

$$\begin{bmatrix} \dfrac{d\bar{x}_1}{d\tau} \\[2em] \dfrac{d\bar{x}_2}{d\tau} \end{bmatrix} = \begin{bmatrix} -1 & -1/\beta \\[1.5em] 0 & \dfrac{\alpha(1-\beta)}{\beta} \end{bmatrix} \begin{bmatrix} \bar{x}_1 \\[2em] \bar{x}_2 \end{bmatrix} \qquad\qquad (1.3\text{-}47)$$

The eigenvalues are

$$\lambda_1 = -1, \qquad \lambda_2 = \alpha(1-\beta)/\beta \qquad\qquad (1.3\text{-}48a,b)$$

The point $(1/\beta, 0)$ is a saddle point and is unstable.

Similarly for the equilibrium point $(1, 1-\beta)$, we write

$$x_1^* = x_1 - 1, \qquad x_2^* = x_2 - 1 + \beta \qquad\qquad (1.3\text{-}49a,b)$$

Proceeding in the usual manner, we deduce that the linearized equations near $(1, 1-\beta)$ can be written as

$$\begin{bmatrix} \dfrac{dx_1^*}{d\tau} \\[2em] \dfrac{dx_2^*}{d\tau} \end{bmatrix} = \begin{bmatrix} -\beta & -1 \\[1.5em] \alpha(1-\beta) & 0 \end{bmatrix} \begin{bmatrix} x_1^* \\[2em] x_2^* \end{bmatrix} \qquad\qquad (1.3\text{-}50)$$

The eigenvalues are

$$\lambda_1 = (-\beta + \sqrt{\beta^2 + 4\alpha\beta - 4\alpha})/2 \qquad\qquad (1.3\text{-}51a)$$

$$\lambda_2 = (-\beta - \sqrt{\beta^2 + 4\alpha\beta - 4\alpha})/2 \qquad\qquad (1.3\text{-}51b)$$

The nature of the point $(1, 1-\beta)$ depends on the sign of $\beta^2 + 4\alpha\beta - 4\alpha$. The equilibrium population of the predator is represented by the dimensionless number $1-\beta$ and this implies that $\beta < 1$. If

$\beta^2 + 4\alpha\beta - 4\alpha$ is positive, λ_1 and λ_2 are real and negative and the point $(1, 1-\beta)$ is an asymptotically stable node. If $\beta^2 + 4\alpha\beta - 4\alpha$ is negative, λ_1 and λ_2 are complex conjugates with negative real parts and the equilibrium is an asymptotically stable focus. In many situations $\alpha \gg \beta$ and, in this case, the point $(1, 1-\beta)$ is an asymptotically stable focus.

The phase portrait for the case $\alpha \gg \beta$ is shown in Figure 1.3-5. The trajectories that start near $(0, 0)$ and $(1/\beta, 0)$ move away from these points and the trajectories that originate near $(1, 1-\beta)$ spiral around it and eventually tend to $(1, 1-\beta)$. That is to say, solutions with initial values approximately equal to $(1, 1-\beta)$ tend to the steady state solution given by

$$x_1 = 1, \qquad x_2 = 1 - \beta \qquad\qquad\qquad (1.3\text{-}52a,b)$$

To describe the trajectories that start far from the equilibrium points, we have to solve the equation

$$\frac{dx_1}{dx_2} = \frac{x_1(1 - \beta x_1 - x_2)}{\alpha x_2 (x_1 - 1)} \qquad\qquad\qquad (1.3\text{-}53)$$

Equation (1.3-53) is deduced from Equations (1.3-45, 29b). No simple analytical solution exists for this equation. The trajectories that originate far from the equilibrium points may spiral to $(1, 1-\beta)$ or may merge into a limit cycle if a periodic solution exists. In the next section, we discuss some criteria which can be used to determine the existence of periodic solutions.

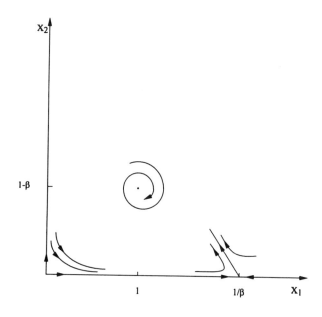

**FIGURE 1.3-5   Phase portrait near the equilibrium points
of a modified Lotka-Volterra equation**

We can investigate the qualitative behavior of Equation (1.3-45) by noting that $\dot{x}_1$ is zero when

$$x_1 = 0, \qquad x_2 = 1 - \beta x_1 \qquad\qquad\qquad\qquad (1.3\text{-}54\text{a, b})$$

These lines are the x_1 **nullclines**. Since x_1 is non-negative, it follows that $\dot{x}_1$ is positive (or negative) if $(1 - \beta x_1 - x_2)$ is positive (or negative). The x_2 nullclines are

$$x_2 = 0, \qquad x_1 = 1 \qquad\qquad\qquad\qquad (1.3\text{-}55\text{a,b})$$

From Equation (1.3-29b), we deduce that $\dot{x}_2$ is positive (or negative) if $x_1 > 1$ $(x_1 < 1)$.

In Figure 1.3-6, we have drawn the nullclines and they divide the first quadrant into four regions. In region I, $\dot{x}_1 > 0$ and $\dot{x}_2 < 0$: the trajectory moves down and to the right, and this is shown by the arrows in the figure. In region II, $\dot{x}_1 > 0$ and $\dot{x}_2 > 0$: the trajectories move upwards and to the right as shown. In region III, $\dot{x}_1 < 0$ and $\dot{x}_2 > 0$: the arrows point upwards and to the left. Finally, in region IV, $\dot{x}_1 < 0$ and $\dot{x}_2 < 0$: the trajectories move downwards and to the left. The trajectories that cross the x_1 (or x_2) nullcline have their tangent lines parallel to the x_2 (or x_1) axis. Combining Figures 1.3-5 and 6, we deduce that the trajectories starting from any point in the first quadrant spiral around and eventually terminate at the point $(1, 1-\beta)$.

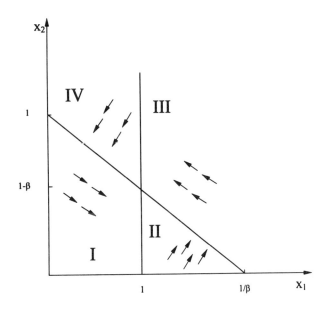

FIGURE 1.3-6   Division of first quadrant by the nullclines and direction of trajectories for a modified Lotka-Volterra equation

Murray (1993) and Lotka (1956) have examined other prey-predator models. We next consider a competition model.

●

Example 1.3-5. **Competition model**

Formulate a competition model and deduce the qualitative behavior of the solutions.

In this case, we consider two species with population N_1 and N_2 competing for limited resources. Assume that the species do not prey on each other. The population of each species grows, in the absence of the other, according to the logistic equation. The competitive effect leads to a decrease in the rate of growth of N_1 and N_2. We consider a simple model and assume dN_1/dt and dN_2/dt to be given by

$$\frac{dN_1}{dt} = a_1 N_1 (1 - N_1/K_1 - b_1 N_2/K_1) \tag{1.3-56a}$$

$$\frac{dN_2}{dt} = a_2 N_2 (1 - N_2/K_2 - b_2 N_1/K_2) \tag{1.3-56b}$$

where a_1, a_2, b_1, b_2, K_1, and K_2 are positive constants.

We introduce the following dimensionless quantities

$$x_1 = N_1/K_1, \qquad x_2 = N_2/K_2, \qquad \tau = a_1 t, \qquad \rho = a_2/a_1,$$
$$\beta_1 = b_1 K_2/K_1, \qquad \beta_2 = b_2 K_1/K_2 \tag{1.3-57a–f}$$

Combining Equations (1.3-56a, b, 57a–f) yields

$$\frac{dx_1}{d\tau} = x_1 (1 - x_1 - \beta_1 x_2) \tag{1.3-58a}$$

$$\frac{dx_2}{d\tau} = \rho x_2 (1 - x_2 - \beta_2 x_1) \tag{1.3-58b}$$

The equilibrium points are $(0, 0)$, $(0, 1)$, $(1, 0)$ and the solution of the set of simultaneous equations

$$x_1 + \beta_1 x_2 = 1, \qquad \beta_2 x_1 + x_2 = 1 \tag{1.3-59a,b}$$

The solution is

$$x_1 = (1 - \beta_1)/(1 - \beta_1 \beta_2), \qquad x_2 = (1 - \beta_2)/(1 - \beta_1 \beta_2) \tag{1.3-60a,b}$$

This solution is relevant only if x_1 and x_2 are non-negative and this implies that β_1 and β_2 must satisfy the following inequalities

$$\beta_1 \geq 1, \quad \beta_2 \geq 1 \quad (\beta_1 \beta_2 > 1) \qquad \text{or} \qquad \beta_1 \leq 1, \quad \beta_2 \leq 1 \quad (\beta_1 \beta_2 < 1)$$

We consider the first case with $\beta_1 = 1.5$ and $\beta_2 = 2$. We have prescribed a numerical value to β_1 and β_2 so as to ease the computation.

The linearized form of Equations (1.3-58a, b), near the origin, is

$$\begin{bmatrix} \dfrac{dx_1}{d\tau} \\[2ex] \dfrac{dx_2}{d\tau} \end{bmatrix} = \begin{bmatrix} 1 & 0 \\[2ex] 0 & \rho \end{bmatrix} \begin{bmatrix} x_1 \\[2ex] x_2 \end{bmatrix} \qquad (1.3\text{-}61)$$

The eigenvalues are 1 and ρ and the origin is an unstable node. To examine the stability of the point $(0, 1)$, we introduce the following transformation

$$\overline{x}_1 = x_1, \qquad \overline{x}_2 = x_2 - 1 \qquad (1.3\text{-}62a,b)$$

Substituting Equations (1.3-62a, b) into Equations (1.3-58a, b) (with $\beta_1 = 1.5$ and $\beta_2 = 2$) and linearizing yields

$$\begin{bmatrix} \dfrac{d\overline{x}_1}{d\tau} \\[2ex] \dfrac{d\overline{x}_2}{d\tau} \end{bmatrix} = \begin{bmatrix} -0.5 & 0 \\[2ex] -2\rho & -\rho \end{bmatrix} \begin{bmatrix} \overline{x}_1 \\[2ex] \overline{x}_2 \end{bmatrix} \qquad (1.3\text{-}63)$$

The eigenvalues are

$$\lambda_1 = -0.5, \qquad \lambda_2 = -\rho \qquad (1.3\text{-}64a,b)$$

The equilibrium point $(0, 1)$ is an asymptotically stable node. Similarly the point $(1, 0)$ is an asymptotically stable node. To analyze the nature of the equilibrium point $(0.25, 0.5)$, we displace the origin to this point by writing

$$x_1^* = x_1 - 0.25, \qquad x_2^* = x_2 - 0.5 \qquad (1.3\text{-}65a,b)$$

The linearized equations can be written as

$$\begin{bmatrix} \dfrac{dx_1^*}{d\tau} \\[2ex] \dfrac{dx_2^*}{d\tau} \end{bmatrix} = \begin{bmatrix} -0.25 & -0.375 \\[2ex] -\rho & -0.5\rho \end{bmatrix} \begin{bmatrix} x_1^* \\[2ex] x_2^* \end{bmatrix} \tag{1.3-66}$$

The eigenvalues are

$$\lambda_{1,2} = [-(0.5\rho + 0.25) \pm \sqrt{(0.5\rho + 0.25)^2 + \rho}\,]/2 \tag{1.3-67}$$

The signs of the eigenvalues are opposite and the equilibrium point is an unstable saddle point. Only two trajectories terminate at this point and only two trajectories depart from this point. The phase diagram near the equilibrium point is shown in Figure 1.3-7.

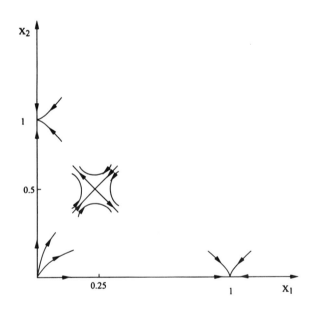

**FIGURE 1.3-7   Phase portrait of a competition model
near the equilibrium points**

The x_1 and x_2 nullclines [Equations (1.3-59a, b)] are shown in Figure 1.3-8 and they divide the positive quadrant into four regions numbered I to IV. In region I, $\dot{x}_1$ and $\dot{x}_2$ are positive; in region II, $\dot{x}_1 > 0$ and $\dot{x}_2 < 0$; in region III, $\dot{x}_1$ and $\dot{x}_2$ are negative; and in region IV, $\dot{x}_1 < 0$ and $\dot{x}_2 > 0$. The directions of the trajectories are indicated by the arrows.

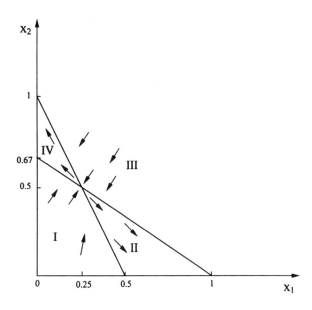

FIGURE 1.3-8   The nullclines and direction of trajectories of a competitive model

Combining Figures 1.3-7 and 8 leads us to deduce that the two trajectories that terminate at the saddle point (0.25, 0.5) are the separatrices that divide the trajectories which stop at (1, 0) from those which stop at (0, 1). The composite portrait is illustrated in Figure 1.3-9.

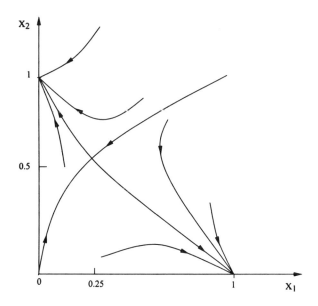

FIGURE 1.3-9   Phase diagram of a competitive model

For the values of β_1 and β_2 we have chosen, there is only one set of initial values that leads to the coexistence of the two species (the separatrices) and all other initial values lead to the ultimate extinction of one of the species.

We now examine the conditions for the coexistence of the two species. The equilibrium point that allows for the coexistence of both species is the one given by Equations (1.3-60a, b) and we denote them by x_{1e} and x_{2e}. To analyze the nature of this equilibrium point, we translate the origin to this point by setting

$$x_1^* = x_1 - x_{1e}, \qquad x_2^* = x_2 - x_{2e} \qquad\qquad (1.3\text{-}68)$$

The linearized equations near (x_{1e}, x_{2e}) can be written as

$$
\begin{bmatrix} \dfrac{dx_1^*}{d\tau} \\[2em] \dfrac{dx_2^*}{d\tau} \end{bmatrix}
=
\begin{bmatrix} -x_{1e} & -\beta_1 x_{1e} \\[1.5em] -\rho\beta_2 x_{2e} & -\rho x_{2e} \end{bmatrix}
\begin{bmatrix} x_1^* \\[2em] x_2^* \end{bmatrix}
\qquad\qquad (1.3\text{-}69)
$$

The eigenvalues are

$$\lambda_{1,2} = \left[-(x_{1e} + \rho x_{2e}) \pm \sqrt{(x_{1e} + \rho x_{2e})^2 - 4\rho\, x_{1e} x_{2e}(1 - \beta_1\beta_2)} \,\right] / 2 \qquad (1.3\text{-}70)$$

We have seen earlier that for (x_{1e}, x_{2e}) to be in the positive quadrant, we require either $\beta_1 > 1$, $\beta_2 > 1$ or $\beta_1 < 1$, $\beta_2 < 1$. From Equation (1.3-70), we deduce that if $\beta_1 > 1$, $\beta_2 > 1$, λ_1 and λ_2 are of opposite sign and (x_{1e}, x_{2e}) is an unstable saddle point; if $\beta_1 < 1$, $\beta_2 < 1$, λ_1 and λ_2 are both negative and (x_{1e}, x_{2e}) is a stable node. We next examine the stability of the other three points when $\beta_1 < 1$ and $\beta_2 < 1$. The linearized equations near the origin [Equation (1.3-61)] are independent of β_1 and β_2 and the origin is an unstable node. The linearized equations in the neighborhood of $(0, 1)$ are obtained by linearizing the combination of Equations (1.3-58a, b, 62a, b). They can be written as

$$
\begin{bmatrix} \dfrac{d\bar{x}_1}{d\tau} \\[2em] \dfrac{d\bar{x}_2}{d\tau} \end{bmatrix}
=
\begin{bmatrix} (1 - \beta_1) & 0 \\[1.5em] -\rho\beta_2 & -\rho \end{bmatrix}
\begin{bmatrix} \bar{x}_1 \\[2em] \bar{x}_2 \end{bmatrix}
\qquad\qquad (1.3\text{-}71)
$$

The eigenvalues are $-\rho$ and $(1 - \beta_1)$ and for $\beta_1 < 1$, the two eigenvalues are of opposite sign and $(0, 1)$ is an unstable saddle point. Similarly for $\beta_2 < 1$, the equilibrium point $(1, 0)$ is an unstable saddle point. The trajectories for the positive quadrant are shown in Figure 1.3-10 for $\beta_1 < 1$ and $\beta_2 < 1$.

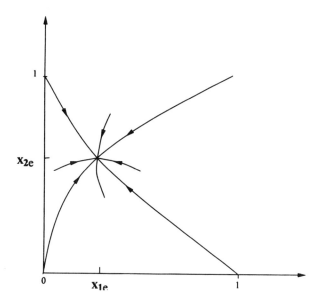

FIGURE 1.3-10   Trajectories in the phase plane for a competition model with $\beta_1 < 1$, $\beta_2 < 1$

The condition for the coexistence of both species is $\beta_1 < 1$ and $\beta_2 < 1$, that is, a coexistence is possible if the competition is not too fierce. If $\beta_1 > 1$ and $\beta_2 > 1$, it generally leads to the extinction of one of the species as shown in the case $\beta_1 = 1.5$ and $\beta_2 = 2$. This may explain the observation that closely related organisms having similar requirements do not occupy the same space. Ecologists refer to this observation as the **competitive exclusion principle** or **Gause's principle** (Odum, 1971).

The next example considers the case of symbiosis (mutualism) where two species cooperate for mutual benefits.

●

Example 1.3-6. **Symbiosis**

Discuss the qualitative behavior of the solutions of the system

$$\frac{dN_1}{dt} = a_1 N_1 (1 - N_1/K_1 + b_1 N_2/K_1) \qquad (1.3\text{-}72a)$$

$$\frac{dN_2}{dt} = a_2 N_2 (1 - N_2/K_2 + b_2 N_1/K_2) \qquad (1.3\text{-}72\text{b})$$

where a_1, a_2, b_1, b_2, K_1, and K_2 are positive constants.

Equations (1.3-72a, b) describe a model where the interaction of two species is to the advantage of both and the sign of b_1 and b_2 is opposite to that in Equations (1.3-56a, b). This model is that of symbiosis (or mutualism) and is of frequent occurrence in nature (Odum, 1971; Murray, 1993).

We non-dimensionalize Equations (1.3-72a, b) by introducing dimensionless variables defined in Equations (1.3-57a–f). Equations (1.3-72a, b) become

$$\frac{dx_1}{d\tau} = x_1 (1 - x_1 + \beta_1 x_2) \qquad (1.3\text{-}73\text{a})$$

$$\frac{dx_2}{d\tau} = \rho x_2 (1 - x_2 + \beta_2 x_1) \qquad (1.3\text{-}73\text{b})$$

The four equilibrium points are $(0, 0)$, $(0, 1)$, $(1, 0)$ and $[x_{1e} = (1+\beta_1)/(1-\beta_1\beta_2)$, $x_{2e} = (1+\beta_2)/(1-\beta_1\beta_2)]$.

The last equilibrium point (x_{1e}, x_{2e}) is relevant only if $1 - \beta_1\beta_2 > 0$. The origin is, as in Example 1.3-5, an unstable node.

To examine the nature of the point $(0, 1)$, we displace the origin to $(0, 1)$ by writing

$$\bar{x}_1 = x_1, \qquad \bar{x}_2 = x_2 - 1 \qquad (1.3\text{-}74\text{a,b})$$

Substituting Equations (1.3-74a, b) into Equations (1.3-73a, b) and linearizing yields

$$\begin{bmatrix} \dfrac{d\bar{x}_1}{d\tau} \\[3mm] \dfrac{d\bar{x}_2}{d\tau} \end{bmatrix} = \begin{bmatrix} 1 + \beta_1 & 0 \\[3mm] \rho\beta_2 & -\rho \end{bmatrix} \begin{bmatrix} \bar{x}_1 \\[3mm] \bar{x}_2 \end{bmatrix} \qquad (1.3\text{-}75)$$

The eigenvalues are $(1+\beta_1)$ and $(-\rho)$ and the equilibrium point is an unstable saddle point. Similarly $(1, 0)$ is an unstable saddle point.

The linearized form of Equations (1.3-73a, b) in the neighborhood of (x_{1e}, x_{2e}) is

$$\begin{bmatrix} \dfrac{dx_1^*}{d\tau} \\[3mm] \dfrac{dx_2^*}{d\tau} \end{bmatrix} = \begin{bmatrix} -x_{1e} & \beta_1 x_{1e} \\[3mm] \rho\beta_2 x_{2e} & -\rho x_{2e} \end{bmatrix} \begin{bmatrix} x_1^* \\[3mm] x_2^* \end{bmatrix} \qquad (1.3\text{-}76\text{a})$$

where

$$x_1^* = x_1 - x_{1e}, \qquad x_2^* = x_2 - x_{2e} \qquad\qquad (1.3\text{-}76\text{b,c})$$

The eigenvalues are given by

$$\lambda_{1,2} = [-(x_{1e} + \rho x_{2e}) \pm \sqrt{(x_{1e} + \rho x_{2e})^2 - 4\rho x_{1e} x_{2e}(1 - \rho\beta_1\beta_2)}]/2 \qquad (1.3\text{-}77)$$

The eigenvalues are real and negative and the point (x_{1e}, x_{2e}) is a stable node. The phase portrait is shown in Figure 1.3-11. If $1 - \beta_1\beta_2 > 0$, x_{1e} and x_{2e} are greater than 1 and the solution tends to (x_{1e}, x_{2e}) as $t \longrightarrow \infty$. This implies that by cooperating, both species can sustain a population greater than its carrying capacity. If $1 - \beta_1\beta_2 < 0$, x_{1e} and x_{2e} are negative and are not relevant. The equilibrium point (x_{1e}, x_{2e}) does not exist and the remaining three equilibrium points are all unstable. In this case, both x_1 and x_2 grow to infinity.

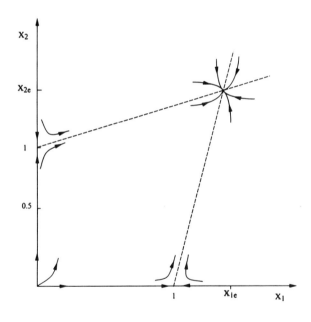

FIGURE 1.3-11   Plot of trajectories for a symbiosis model
- - - : nullclines

All the models considered so far are described by kinetic equations which can be written as

$$\frac{dN_i}{dt} = N_i F_i (N_1, N_2, ...), \qquad i = 1, 2, ... \qquad (1.3\text{-}78)$$

Note that it is assumed that the origin is an equilibrium point and if N_i is zero initially, $\dot{N}_i$ is zero and this implies that N_i is zero at all subsequent times. However, the origin is not necessarily an asymptotically stable point and any perturbation near the origin may lead to an increase in the values of N_i. Here, we have considered only two species and we have been able to describe the evolution of N_i in the phase plane. The analysis can be extended to n species (n > 2) and the trajectories have to be drawn in n-dimensional space. This increase of dimensionality can lead to an increase in complexity. Consider the case of three species. The matrix for the linearized equations near an equilibrium point is a 3×3 matrix and its characteristic equation is cubic with eigenvalues λ_1, λ_2, and λ_3. It is possible for λ_1 to be real and positive, and for λ_2 and λ_3 to be complex conjugates with negative real parts. We now transform the matrix to its canonical form and introduce the coordinates (y_1, y_2, y_3) as in the two-dimensional case [Equation (1.2-10)]. The solution y_1, associated with λ_1, grows exponentially to infinity while y_2 and y_3 spiral around towards the equilibrium point. This equilibrium point is a **saddle focus** with a destabilizing trend coexisting with a stabilizing one. For the linear system, the trajectory eventually tends to infinity along the y_1-axis. For the non-linear system, the trajectory might not tend to infinity (Example 1.3-2) and the trajectory is turned around and is confined to a finite space. If there is no stable equilibrium point to which the trajectory can tend to, the trajectory wanders in the three-dimensional space, sometimes approaching the equilibrium point and at other times departing from the equilibrium point. This is **deterministic chaos** which will be examined in Chapter 5. In two dimensions, chaos does not occur.

Equation (1.3-78) can be used to describe not only population dynamics but also chemical reactions and many other processes. We close this section with an example of a chemical reaction.

Example 1.3-7. **Chemical reaction**

The Brusselator was proposed by Prigogine and Lefever (1968) and describes the following set of reactions

$$A \xrightarrow{k_1} X, \quad B + X \xrightarrow{k_2} Y + C, \quad 2X + Y \xrightarrow{k_3} 3X, \quad X \xrightarrow{k_4} D$$

where $k_1, ... , k_4$ are rate constants. The concentrations of A, B, C, and D are kept constant. We wish to deduce the rate equations for the intermediary products X and Y and to discuss the qualitative properties of the solutions of the resulting equations.

Applying the law of conservation of mass, one obtains

$$\frac{d[X]}{dt} = k_1[A] - k_2[B][X] + k_3[X]^2[Y] - k_4[X] \tag{1.3-79a}$$

$$\frac{d[Y]}{dt} = k_2[B][X] - k_3[X]^2[Y] \tag{1.3-79b}$$

We non-dimensionalize Equations (1.3-79a, b) by introducing the following dimensionless quantities

$$\tau = k_4 t, \qquad x_1 = k_4 [X]/k_1 [A], \qquad x_2 = k_4 [Y]/k_1 [A]$$

$$\alpha = k_3 (k_1 [A])^2 / k_4^3, \qquad \beta = k_2 [B]/k_4$$
(1.3-80a–e)

Combining Equations (1.3-79a – 80e) yields

$$\frac{dx_1}{d\tau} = 1 - (1 + \beta) x_1 + \alpha x_1^2 x_2$$
(1.3-81a)

$$\frac{dx_2}{d\tau} = \beta x_1 - \alpha x_1^2 x_2$$
(1.3-81b)

The equilibrium point is $(1, \beta/\alpha)$.

The linearized equations near $(1, \beta/\alpha)$ can be written as

$$\begin{bmatrix} \dfrac{dx_1^*}{d\tau} \\[4mm] \dfrac{dx_2^*}{d\tau} \end{bmatrix} = \begin{bmatrix} \beta - 1 & \alpha \\[4mm] -\beta & -\alpha \end{bmatrix} \begin{bmatrix} x_1^* \\[4mm] x_2^* \end{bmatrix}$$
(1.3-82a)

where $x_1^* = x_1 - 1, \qquad x_2^* = x_2 - \beta/\alpha$
(1.3-82b,c)

The eigenvalues are

$$\lambda_{1,2} = [-(1 + \alpha - \beta) \pm \sqrt{(1 + \alpha - \beta)^2 - 4\alpha}]/2$$
(1.3-83a)

$$= [-(1 + \alpha - \beta) \pm \sqrt{(1 + \alpha + 2\sqrt{\alpha} - \beta)(1 + \alpha - 2\sqrt{\alpha} - \beta)}]/2$$
(1.3-83b)

From Equation (1.3-83b), we deduce that

(i)   if $(1 + \alpha - 2\sqrt{\alpha}) < \beta < (1 + \alpha + 2\sqrt{\alpha})$, the eigenvalues are complex. Further, if $\beta = 1 + \alpha$, the eigenvalues are purely imaginary and the equilibrium point is a center; if $(1 + \alpha) > \beta$, the equilibrium point is a stable focus; and if $(1 + \alpha) < \beta$, the equilibrium point is an unstable spiral.

(ii)   if $(1 + \alpha - 2\sqrt{\alpha}) > \beta$, the eigenvalues are real and negative and the equilibrium point is a stable node.

(iii)   if $(1 + \alpha + 2\sqrt{\alpha}) < \beta$, the eigenvalues are real and positive and the equilibrium point is an unstable node.

The linear analysis shows that for a fixed value of α and for all values of $\beta < (1 + \alpha)$, the equilibrium point is stable, and for all values of $\beta > (1 + \alpha)$, the equilibrium point is unstable. The point $\beta = (1 + \alpha)$ separates the stability from the instability regions and is a **bifurcation point**. Bifurcation points are defined and discussed in Chapter 4.

It has been shown that the Brusselator does oscillate (Example 4.3-3 and Lefever et al., 1988). The next section consider the possibility of oscillation in two-dimensional dynamical systems.

1.4   EXISTENCE AND NON-EXISTENCE OF PERIODIC SOLUTIONS

In Section 1.3, we deduce the qualitative properties of the solutions of Equations (1.2-1a, b) by examining the solutions near the equilibrium points. In this section, we shall state conditions for the existence or non-existence of periodic solutions in the whole phase plane. We start by defining the **Poincaré index**.

Poincaré Index

Let C be a smooth closed curve that does not pass through an equilibrium point of Equations (1.2-1a, b). At each point of C, we can associate an angle ϕ defined by

$$\phi = \tan^{-1}(f_2/f_1) \tag{1.4-1}$$

On going around C, ϕ changes continuously and on returning to the original starting point, the value of ϕ has changed by a multiple of 2π.

The Poincaré index I is defined by

$$I = \frac{1}{2\pi} \int_C d\phi = \frac{1}{2\pi} \int_C d\,[\tan^{-1}(f_2/f_1)] \tag{1.4-2}$$

The evaluation of the integral is in the anti-clockwise direction. We recall that

$$d\,(\tan^{-1}x) = [1/(1 + x^2)]\,dx \tag{1.4-3}$$

It follows that

$$d\,[\tan^{-1}(f_2/f_1)] = [1/(1 + f_2^2/f_1^2)]\,[(f_1\,df_2 - f_2\,df_1)/f_1^2] \tag{1.4-4a}$$

$$= (f_1\,df_2 - f_2\,df_1)/(f_1^2 + f_2^2) \tag{1.4-4b}$$

Substituting Equation (1.4-4b) into Equation (1.4-2) yields

$$I = \frac{1}{2\pi} \int_C (f_1 df_2 - f_2 df_1) / (f_1^2 + f_2^2) \qquad (1.4-5)$$

The df_i $(i = 1, 2)$ can be written as

$$df_i = \frac{\partial f_i}{\partial x_1} dx_1 + \frac{\partial f_i}{\partial x_2} dx_2 \qquad (1.4-6)$$

In some cases, it is easier to work in polar coordinates, as shown in the next example.

Example 1.4-1. Poincaré index

Calculate the Poincaré index associated with the unit circle centered at the origin of the following system

$$\frac{dx_1}{dt} = 2x_1^2 - 1, \qquad \frac{dx_2}{dt} = 2x_1 x_2 \qquad (1.4-7a,b)$$

Transforming to polar coordinates, we write

$$x_1 = \cos\theta, \qquad x_2 = \sin\theta \qquad (1.4-8a,b)$$

$$f_1 = 2x_1^2 - 1 = 2\cos^2\theta - 1 = \cos 2\theta \qquad (1.4-9a,b,c)$$

$$f_2 = 2x_1 x_2 = 2\cos\theta \sin\theta = \sin 2\theta \qquad (1.4-10a,b,c)$$

From Equation (1.4-1), we deduce that

$$\phi = 2\theta \qquad (1.4-11)$$

On increasing θ by 2π (going once around C), ϕ is increased by 4π.

We can compute df_i from Equations (1.4-6, 9a–10c) and obtain

$$df_1 = -2\sin 2\theta \, d\theta, \qquad df_2 = 2\cos 2\theta \, d\theta \qquad (1.4-12a,b)$$

Substituting Equations (1.4-9c, 10c, 12a, b) into Equation (1.4-5) yields

$$I = 0 \qquad (1.4-13)$$

●

We next state the value of I for the system given by Equations (1.2-1a, b) under various conditions.

(i)   The index I is zero if, on and inside C, the system has no equilibrium point, f_1 and f_2 have continuous first derivatives.

(ii)   If C_1 is a closed curve inside C and there is no equilibrium point in the region enclosed by C and C_1, the index associated with C is equal to that associated with C_1.

This implies that the index is independent of C and depends on the presence of equilibrium points inside C. It is more appropriate to state the values of I associated with equilibrium points.

(iii)   The index of a node, focus, and center is 1, that of a saddle is –1.

(iv)   If C encloses n equilibrium points and the index of the jth equilibrium point is I_j, the index I of C is

$$I = \sum_{j=1}^{n} I_j \qquad\qquad (1.4\text{-}14)$$

That is to say, the index is additive.

The proof of statements (i) to (iv) are given in Cesari (1971) and Jordan and Smith (1977).

We can now deduce that if C is a periodic orbit, C must enclose an equilibrium point and its index must be one. This is a necessary but not a sufficient condition. If its index is not +1, C cannot be a periodic orbit. This provides us with a criterion to rule out the existence of a periodic solution, as shown next.

Example 1.4-2. **Poincaré index and periodic solutions**

Calculate the index of the equilibrium point (0, 0) of

$$\frac{dx_1}{dt} = x_2^3, \qquad\qquad \frac{dx_2}{dt} = x_1^3 \qquad\qquad (1.4\text{-}15a,b)$$

The origin is the only equilibrium point. We choose C to be a square of length 2, as shown in Figure 1.4-1.

Combining Equations (1.4-5, 6, 15a, b) yields

$$I = \frac{1}{2\pi} \int_C \frac{3\,(x_1^2 x_2^3\,dx_1 - x_1^3 x_2^2\,dx_2)}{(x_1^6 + x_2^6)} \qquad\qquad (1.4\text{-}16)$$

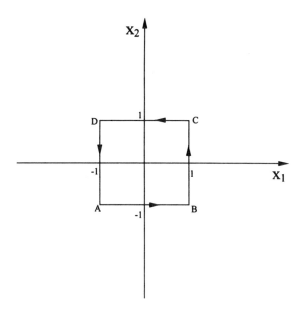

FIGURE 1.4-1   Closed curve C for the evaluation of I

We integrate in the anti-clockwise direction, along AB, BC, CD, and DA.

Along AB, $x_2 = -1$ and $dx_2 = 0$. We need to evaluate [see Equation (1.4-16)]

$$\int_{-1}^{1} \frac{-3x_1^2\, dx_1}{1 + x_1^6} \;=\; -\int_{-1}^{1} \frac{dy}{1 + y^2} \;=\; -\left[\tan^{-1}y\right]_{-1}^{1} \;=\; -\frac{\pi}{2} \qquad (1.4\text{-}17a,b,c)$$

Similarly we can evaluate the integrals along BC, CD, and DA and each is equal to $-\pi/2$. We deduce that I is (-1) and the system has no periodic solution.

●

Several other criteria have been established to show the non-existence of periodic solutions and we next state two of them.

Bendixson's Negative Criterion

If $(\partial f_1/\partial x_1 + \partial f_2/\partial x_2)$ does not change sign in a certain region of the phase plane, the system given by Equations (1.2-1a, b) does not have a closed trajectory in that region.

The proof of this theorem can be found in Chan Man Fong et al. (1997) and Jordan and Smith (1977).

Dulac's Criterion

If $\rho(x_1, x_2)$ is a continuously differentiable function and $\left[\dfrac{\partial}{\partial x_1}(\rho f_1) + \dfrac{\partial}{\partial x_2}(\rho f_2)\right]$ is of one sign in a simply connected region, then the system defined by Equations (1.2-1a, b) does not have a periodic trajectory in the region.

If $\rho = 1$, Dulac's criterion reduces to Bendixson's criterion.

Example 1.4-3. **Non-existence of a periodic solution**

Show that the system

$$\dot{x}_1 = x_1(1 - x_1 - \beta_2 x_2) \tag{1.4-18a}$$

$$\dot{x}_2 = \alpha x_2(1 - x_2 - \beta_1 x_1) \tag{1.4-18b}$$

has no periodic solution.

Equations (1.4-18a, b) are the Lotka-Volterra equations (Murray, 1993). Here, unlike in Example 1.3-4, both species, in the absence of the other, have a logistic growth.

Applying Bendixson's criterion yields

$$\frac{\partial f_1}{\partial x_1} + \frac{\partial f_2}{\partial x_2} = (1 - 2x_1 - \beta_2 x_2) + (\alpha - 2\alpha x_2 - \alpha\beta_1 x_1) \tag{1.4-19a}$$

$$= (1 + \alpha) - [x_1(2 + \alpha\beta_1) + x_2(2\alpha + \beta_2)] \tag{1.4-19b}$$

No definite conclusion can be drawn since the sign of $(\partial f_1/\partial x_1 + \partial f_2/\partial x_2)$ may change.

Applying Dulac's criterion with $\rho = 1/(x_1 x_2)$ yields

$$\frac{\partial}{\partial x_1}(\rho f_1) + \frac{\partial}{\partial x_2}(\rho f_2) = -\left(\frac{1}{x_2} + \frac{\alpha}{x_1}\right) \tag{1.4-20}$$

The quantities x_1 and x_2 are positive and in the positive quadrant (excluding the origin where ρ is singular) the sign of $\dfrac{\partial}{\partial x_1}(\rho f_1) + \dfrac{\partial}{\partial x_2}(\rho f_2)$ is always negative and the system has no periodic solution. The origin is an unstable node.

●

The next theorem states the conditions that guarantee the existence of a closed trajectory (periodic solution).

Poincaré-Bendixson's Theorem

Let f_1 and f_2 in Equations (1.2-1a, b) have continuous first order partial derivatives in a domain D of the (x_1, x_2)-plane and let D_1 be a bounded domain of D. Let R be the region that consists of D_1 and its boundary and we assume that the system given by Equations (1.2-1a, b) has no equilibrium point in R. If C $[= (x_1(t), x_2(t))]$ is a trajectory of the system lying in R for some t_0 and remains in R for all $t \geq t_0$, C is either a closed trajectory or a spiral going toward a closed trajectory as $t \longrightarrow \infty$. In either case, the system has a periodic solution. The proof is given in Cesari (1971).

Note that if there is a closed trajectory in R, this closed trajectory must enclose an equilibrium point with index 1, and this equilibrium point cannot be in R. It follows that R is not a simply connected region, that is, it must have a hole. The next example illustrates this.

Example 1.4-4. **Existence of a periodic solution**

Show that the system

$$\dot{x}_1 = -x_2 + x_1 (1 - x_1^2 - x_2^2) \tag{1.4-21a}$$

$$\dot{x}_2 = x_1 + x_2 (1 - x_1^2 - x_2^2) \tag{1.4-21b}$$

has a periodic solution.

The origin is the only equilibrium point.

Introducing polar coordinates (r, θ) and combining Equations (1.2-28a, b, 4-21a, b) yields

$$\dot{r} = r(1 - r^2), \qquad \dot{\theta} = 1 \tag{1.4-22a,b}$$

Consider the region R enclosed by the circles $r = \varepsilon (< 1)$ and $r = K (> 1)$. There is no equilibrium point in R. If $r < 1$, we deduce from Equation (1.4-22a) that r increases with time and any trajectory starting near $r = \varepsilon$ moves away from the circle. If $r > 1$, r decreases with time and any trajectory near $r = K$ moves away from this boundary. Thus any trajectory that starts in R stays in R and by the Poincaré-Bendixson's theory, there is a closed trajectory of the system in R.

We can solve Equations (1.4-22a, b) and the solution is

$$r^2 = r_0^2 / [r_0^2 + (1 - r_0^2) e^{-2t}], \qquad \theta = \theta_0 + t \tag{1.4-23a,b}$$

where r_0 and θ_0 are the initial values of r and θ respectively.

The limit cycle (closed trajectory) is the circle $r = 1$. If $r_0 < 1$, the trajectory spirals outwards to $r = 1$, and if $r_0 > 1$, the trajectory spirals inwards to $r = 1$. The limit cycle ($r = 1$) encloses the origin which is a center with index 1. Note that R does not include the origin. If we apply Bendixson's criterion, we obtain

$$\frac{\partial f_1}{\partial x_1} + \frac{\partial f_2}{\partial x_2} = 2 [1 - 2(x_1^2 + x_2^2)] \tag{1.4-24}$$

In the region R, $(\partial f_1/\partial x_1 + \partial f_2/\partial x_2)$ changes sign and we deduce that there is a possibility of a closed trajectory but we are not guaranteed of its existence. The Poincaré-Bendixson's theorem confirms the existence of a periodic solution.

●

Example 1.4-5. **Possibility of a periodic solution**

Discuss the possibility of a periodic solution of the equation

$$\ddot{x} - \mu(1 - x^2)\dot{x} + x = 0 \tag{1.4-25}$$

where μ is a positive constant.

Equation (1.4-25) can be written as

$$x_1 = x \tag{1.4-26a}$$

$$\dot{x}_1 = x_2 \tag{1.4-26b}$$

$$\dot{x}_2 = -x_1 + \mu(1 - x_1^2)x_2 \tag{1.4-26c}$$

The origin is the only equilibrium point. Near the origin, the corresponding linear system is

$$\begin{bmatrix} \dot{x}_1 \\ \dot{x}_2 \end{bmatrix} = \begin{bmatrix} 0 & 1 \\ -1 & \mu \end{bmatrix} \begin{bmatrix} x_1 \\ x_2 \end{bmatrix} \tag{1.4-27}$$

The eigenvalues are

$$\lambda_{1,2} = [\mu \pm \sqrt{\mu^2 - 4}]/2 \tag{1.4-28}$$

The origin is an unstable spiral if $0 < \mu < 2$ and an unstable node if $\mu \geq 2$. Note that if $\mu = 0$, the origin is a center and the solution is periodic.

To apply Bendixson's theorem, we compute

$$\frac{\partial}{\partial x_1}(x_2) + \frac{\partial}{\partial x_2}[-x_1 + \mu(1 - x_1^2)x_2] = \mu(1 - x_1^2) \tag{1.4-29}$$

Since $(1 - x_1^2)$ can change sign, we conclude that a periodic solution is possible. To confirm its existence, we apply the Poincaré-Bendixson theorem. On transforming to polar coordinate, Equations (1.4-26b, c) become (see Example 1.4-4)

$$\dot{r} = \mu(1 - r^2\cos^2\theta)\, r\sin^2\theta \tag{1.4-30a}$$

$$\dot{\theta} = -1 + \mu(1 - r^2\cos^2\theta)\sin\theta\cos\theta \tag{1.4-30b}$$

Consider the region R bounded by the circles $r = \varepsilon (<1)$ and $r = K (> 1)$. Near the circle $r = \varepsilon$, $\dot{r}$ is positive or zero (along the axis $\theta = 0, \pi$) and the trajectories starting near $r = \varepsilon$ move away from the circle. Near the circle $r = K$, $\dot{r}$ is negative except along the axes $\theta = 0, \pi/2, \pi, 3\pi/2$. Along $\theta = 0, \pi$, $\dot{r}$ is zero, but along $\theta = \pi/2, 3\pi/2$, $\dot{r}$ is positive. The trajectories starting near $r = K$ move inwards in R except along $\theta = \pi/2, 3\pi/2$ where they move away from R. For this choice of R, we cannot apply the Poincaré-Bendixson theorem. Other theorem have been proved to establish the existence of periodic solutions and this is considered next.

●

Equation (1.4-25) is the Van der Pol equation and it arises in the theory of circuits containing a triode valve. Since the Second World War, this equation has stimulated many investigations in the theory of non-linear differential equations. Here, we state only two theorems; proofs and further details are given in Cesari (1971), Jordan and Smith (1987), and Minorsky (1962).

Theorem 1

The equation

$$\ddot{x} + f(x)\dot{x} + g(x) = 0 \tag{1.4-31}$$

has a unique periodic solution if

(i) f and g are continuous,

(ii) $F(x) = \displaystyle\int_0^x f(\xi)\, d\xi \tag{1.4-32}$

   is an odd function,

(iii) $F(x)$ is zero only at $x = 0$ and $x = \pm a$,

(iv) $F(x) \longrightarrow \infty$ monotonicly for $x > a$,

(v) $g(x)$ is an odd function and $g(x) > 0$ for $x > 0$.

Theorem 2

The equation

$$\ddot{x} + f(x, \dot{x})\, \dot{x} + g(x) = 0 \tag{1.4-33}$$

has at least one periodic solution if

(i) f and g are continuous,

(ii) $f(x, \dot{x}) > 0$ when $(x^2 + \dot{x}^2) > a^2$,

(iii) $f(x, \dot{x}) < 0$ in the neighborhood of and at the origin,

(iv) $g(x) > 0$ when $x > 0$, $g(x) < 0$ when $x < 0$, and $g(0) = 0$,

(v) $G(x) = \displaystyle\int_0^x g(\xi)\, d\xi$

         tends to infinity as $x \longrightarrow \infty$.

Applying Theorem 1 to Equation (1.4-25), we deduce that it has a periodic solution and $a = \sqrt{3}$.

We next examine a linear system with periodic coefficients.

1.5   FLOQUET'S THEOREM

Consider the n-dimensional, linear, and non-autonomous system defined by

$$\dot{\underline{x}} = \underline{\underline{A}}(t)\, \underline{x} \tag{1.5-1}$$

where $\underline{A}(t)$ is a $(n \times n)$ matrix whose elements are continuous functions of t.

We determine under what conditions the system has a periodic solution if $\underline{A}(t)$ is periodic (i.e. all elements of $\underline{A}$ are periodic functions). Before doing this, we consider a scalar equation and review the solutions of linear systems of differential equations.

Example 1.5-1. **Scalar differential equations**

Solve the following equations

(a) $\dot{x} = (\tan t)\, x$ (1.5-2a)

(b) $\dot{x} = (\cos^2 t)\, x$ (1.5-2b)

(a)   The integrating factor of Equation (1.5-2a) is $\cos t$ and on multiplying by the integrating factor, we obtain

$$\frac{d}{dt}(x \cos t) = 0 \qquad (1.5\text{-}3)$$

The solution is

$$x \cos t = C_0 \qquad (1.5\text{-}4)$$

where C_0 is a constant.

The solution is periodic.

(b)   Equation (1.5-2b) can be written as

$$\frac{dx}{x} = \cos^2 t \; dt \qquad (1.5\text{-}5)$$

Integrating yields

$$x = C_0 \exp[t/2 + (1/4)\sin 2t] \qquad (1.5\text{-}6)$$

where C_0 is a constant.

In this case, the solution is not periodic.

We note that an equation with periodic coefficients does not necessarily have periodic solutions.

●

Let $\underline{\phi}_i(t)$ be a solution of Equation (1.5-1), that is to say

$$\underline{\dot{\phi}}_i = \underline{\underline{A}}\, \underline{\phi}_i \qquad (1.5\text{-}7)$$

The vector $\underline{\phi}_i$ has n elements $[\phi_{1i}, \phi_{2i}, \ldots, \phi_{ni}]$ and there are n vectors $\underline{\phi}_i$ $(i = 1, \ldots, n)$ that satisfy Equation (1.5-7). If the n vectors are linearly independent, the general solution $\underline{x}$ of Equation (1.5-1) is

$$\underline{x} = \sum_{i=1}^{n} c_i \, \underline{\phi}_i \tag{1.5-8}$$

where c_i are constants.

The $\underline{\phi}_i$ $(i = 1, \ldots, n)$ are **fundamental solutions** and a **fundamental matrix** is defined by

$$\underline{\underline{\Phi}} = \begin{bmatrix} \underline{\phi}_1 & \underline{\phi}_2 & \cdots & \underline{\phi}_n \end{bmatrix} = \begin{bmatrix} \phi_{11} & \phi_{12} & \cdots & \phi_{1n} \\ \phi_{21} & \phi_{22} & \cdots & \phi_{2n} \\ \vdots & \vdots & & \vdots \\ \phi_{n1} & \phi_{n2} & \cdots & \phi_{nn} \end{bmatrix} \tag{1.5-9a,b}$$

The **Wronskian** W is the determinant $|\underline{\underline{\Phi}}|$. If the $\underline{\phi}_i$ are linearly independent, W is non-zero.

By definition, $\underline{\underline{\Phi}}$ satisfies Equation (1.5-1), that is

$$\dot{\underline{\underline{\Phi}}} = \underline{\underline{A}} \, \underline{\underline{\Phi}} \tag{1.5-10}$$

The general solution $\underline{x}$ [Equation (1.5-8)] is given by

$$\underline{x} = \underline{\underline{\Phi}} \, \underline{c} \tag{1.5-11}$$

where $\underline{c}$ is a constant vector.

Suppose the initial conditions are

$$\underline{x}(0) = \underline{x}_0 \tag{1.5-12}$$

Substituting Equation (1.5-12) into Equation (1.5-11), we deduce

$$\underline{c} = \left[\underline{\underline{\Phi}}(0) \right]^{-1} \underline{x}_0 \tag{1.5-13}$$

It follows that the general solution is

$$\underline{x} = \underline{\underline{\Phi}}(t)\left[\underline{\underline{\Phi}}(0)\right]^{-1}\underline{x}_0 \tag{1.5-14}$$

The relation between $W(t)$ and $W(0)$ is

$$W(t) = W(0)\exp\left[\int_0^t \text{tr}\,\underline{\underline{A}}(s)\,ds\right] \tag{1.5-15}$$

Equation (1.5-15) is Liouville's formula and the proof is given in Jordan and Smith (1987).

Note that if $\underline{\phi}_1$ and $\underline{\phi}_2$ are two fundamental solutions, $\underline{\phi}_1$ and $(\alpha_1\underline{\phi}_1 + \alpha_2\underline{\phi}_2)$, where α_1 and α_2 are constants, are also fundamental solutions. If $\underline{\underline{\Phi}}_1$ and $\underline{\underline{\Phi}}_2$ are two fundamental matrices, they are related by

$$\underline{\underline{\Phi}}_1(t) = \underline{\underline{\Phi}}_2(t)\,\underline{\underline{C}} \tag{1.5-16}$$

where $\underline{\underline{C}}$ is a constant matrix.

Equation (1.5-16) follows from the fact that $\underline{\underline{\Phi}}_2$ is a linear combination of the columns of $\underline{\underline{\Phi}}_1$.

We can now state and prove Floquet's theorem.

Floquet's theorem

If $\underline{\underline{A}}(t)$ in Equation (1.5-1) is periodic and of period T $[\underline{\underline{A}}(t+T) = \underline{\underline{A}}(t)]$, there is at least one solution $\underline{x}(t)$ such that

$$\underline{x}(t+T) = \lambda\,\underline{x}(t) \tag{1.5-17}$$

where λ is a constant.

Proof. Let $\underline{\underline{\Phi}}(t)$ be a fundamental matrix of Equation (1.5-1). Since $\underline{\underline{A}}(t+T) = \underline{\underline{A}}(t)$, $\underline{\underline{\Phi}}(t+T)$ also satisfies Equation (1.5-1). From Equation (1.5-15), we deduce that $|\underline{\underline{\Phi}}(t+T)|$ is non-zero and $\underline{\underline{\Phi}}(t+T)$ is also a fundamental solution. These two fundamental matrices are related by [Equation (1.5-16)]

$$\underline{\underline{\Phi}}(t+T) = \underline{\underline{\Phi}}(t)\,\underline{\underline{C}} \tag{1.5-18}$$

Let λ and $\underline{v}$ be an eigenvalue and an eigenvector of $\underline{\underline{C}}$ respectively. By definition

$$\underline{\underline{C}}\,\underline{v} = \lambda\,\underline{v} \tag{1.5-19}$$

Consider the solution $\underline{x}(t)$ given by

$$\underline{x}(t) = \underline{\underline{\Phi}}\,\underline{v} \tag{1.5-20}$$

At time $t + T$, we have, using Equations (1.5-18, 19, 20),

$$\underline{x}(t + T) = \underline{\underline{\Phi}}(t + T)\,\underline{v} = \underline{\underline{\Phi}}(t)\,\underline{\underline{C}}\,\underline{v} = \lambda\,\underline{x}(t) \tag{1.5-21a,b,c}$$

This completes the proof.

In general, $\underline{\underline{C}}$ has n eigenvalues $\lambda_1, \lambda_2, \ldots, \lambda_n$ and these eigenvalues are the **characteristic factors** or **multipliers** of Equation (1.5-1). The λ_s $(s = 1, \ldots, n)$ can also be written as

$$\lambda_s = \exp(r_s T), \qquad s = 1, \ldots, n \tag{1.5-22}$$

The numbers r_s are the **characteristic exponents** of the system. Note that $\exp(2\pi i)$ is one and it follows that λ_s are determined up to multiples of $2\pi i/T$.

Equation (1.5-1), with $\underline{\underline{A}}(t)$ periodic of period T, has a periodic solution of period T if one of the λ_s is one (or one of the r_s is zero). If λ_s is (-1), we have a periodic solution of period $2T$.

To apply Floquet's theorem, we need to know $\underline{\underline{\Phi}}$ so as to determine $\underline{\underline{C}}$ and this is not always possible. From Equations (1.5-15, 18), we deduce

$$W(T) = W(0)\det\underline{\underline{C}} \tag{1.5-23a}$$

$$\det\underline{\underline{C}} = \exp\int_0^T \mathrm{tr}\,\underline{\underline{A}}(s)\,ds \tag{1.5-23b}$$

The product of $\lambda_1\,\lambda_2\,\ldots\,\lambda_n$ is given by

$$\lambda_1\,\lambda_2\,\ldots\,\lambda_n = \det\underline{\underline{C}} = \exp\int_0^T \mathrm{tr}\,\underline{\underline{A}}(s)\,ds \tag{1.5-24a,b}$$

Equation (1.5-24b) is sometimes useful to determine if one of the eigenvalues is unity (see Example 1.5-2).

The next few examples illustrate the use of Floquet's theorem.

Example 1.5-2. Fundamental matrix

Find a fundamental matrix for the system

$$\dot{\underline{x}} = \underline{A}(t)\,\underline{x} \tag{1.5-25a}$$

$$\underline{A} = \begin{bmatrix} 1 & 1 \\ 0 & \dfrac{\cos t + \sin t}{2 + \sin t - \cos t} \end{bmatrix} \tag{1.5-25b}$$

Determine (a) the characteristic factors, (b) the characteristic exponents, (c) the existence of periodic solutions.

The equation for x_2 is

$$\frac{dx_2}{dt} = \left(\frac{\cos t + \sin t}{2 + \sin t - \cos t}\right) x_2 \tag{1.5-26a}$$

or
$$\frac{dx_2}{x_2} = \left(\frac{\cos t + \sin t}{2 + \sin t - \cos t}\right) dt \tag{1.5-26b}$$

Equation (1.5-26b) can be integrated to yield

$$\ln x_2 = \ln(2 + \sin t - \cos t) + \ln c_1 \tag{1.5-27a}$$

or
$$x_2 = c_1(2 + \sin t - \cos t) \tag{1.5-27b}$$

where c_1 is a constant.

The equation for x_1 is

$$\frac{dx_1}{dt} = x_1 + x_2 = x_1 + c_1(2 + \sin t - \cos t) \tag{1.5-28a,b}$$

Equation (1.5-28b) is a first order equation and its solution is

$$x_1 = -c_1(2 + \sin t) + c_2 e^t \tag{1.5-29}$$

where c_2 is a constant.

A fundamental matrix is

$$\underline{\Phi}(t) = \begin{bmatrix} -(2 + \sin t) & e^t \\ (2 + \sin t - \cos t) & 0 \end{bmatrix} \tag{1.5-30}$$

The coefficient matrix $\underline{A}$ is of period 2π. We evaluate $\underline{\Phi}(0)$ and $\underline{\Phi}(2\pi)$ and they are

$$\underline{\underline{\Phi}}(0) = \begin{bmatrix} -2 & 1 \\ 1 & 0 \end{bmatrix} \qquad (1.5\text{-}31\text{a})$$

$$\underline{\underline{\Phi}}(2\pi) = \begin{bmatrix} -2 & e^{2\pi} \\ 1 & 0 \end{bmatrix} \qquad (1.5\text{-}31\text{b})$$

The constant matrix $\underline{\underline{C}}$ [Equation (1.5-18)] is given by

$$\underline{\underline{C}} = [\underline{\underline{\Phi}}(0)]^{-1}\,\underline{\underline{\Phi}}(2\pi) = \begin{bmatrix} 1 & 0 \\ 0 & e^{2\pi} \end{bmatrix} \qquad (1.5\text{-}32\text{a,b})$$

The eigenvalues of $\underline{\underline{C}}$ are given by

$$\begin{vmatrix} 1-\lambda & 0 \\ 0 & e^{2\pi}-\lambda \end{vmatrix} = 0 \qquad (1.5\text{-}33)$$

The solutions are

$$\lambda_1 = 1, \qquad \lambda_2 = e^{2\pi} \qquad (1.5\text{-}34\text{a,b})$$

The characteristic factors are 1 and $e^{2\pi}$.

The characteristic exponents are 0 and 1.

Since one of the eigenvalues is one (corresponding to a characteristic exponent being zero), then the system has a periodic solution. This periodic solution corresponds to $c_1 \neq 0$ and $c_2 = 0$ and can be written as

$$\underline{\phi}_1 = \begin{bmatrix} -(2 + \sin t) \\ 2 + \sin t - \cos t \end{bmatrix} \qquad (1.5\text{-}35)$$

Example 1.5-3. **Hill's equation**

Consider the second order

$$\ddot{x} + p(t)\,x = 0 \qquad (1.5\text{-}36)$$

where $p(t)$ is periodic with period π. Choose a fundamental matrix $\underline{\Phi}(t)$ such that

$$\underline{\Phi}(0) = \underline{I} \tag{1.5-37}$$

where $\underline{I}$ is the identity matrix.

Determine the conditions on $\underline{\Phi}(\pi)$ such that Equation (1.5-36) has periodic solutions.

Equation (1.5-36) is Hill's equation and occurs frequently in vibrations and in electrical circuits (Minorsky, 1962). In the Mathieu equation, $p(t)$ is given by

$$p(t) = \delta + \varepsilon \cos 2t \tag{1.5-38}$$

where δ and ε are constants.

Equation (1.5-36) can be written as

$$\begin{bmatrix} \dot{x}_1 \\ \dot{x}_2 \end{bmatrix} = \begin{bmatrix} 0 & 1 \\ -p & 0 \end{bmatrix} \begin{bmatrix} x_1 \\ x_2 \end{bmatrix} \tag{1.5-39}$$

where $x_1 = x$.

From Equation (1.5-18), we deduce that

$$\underline{C} = \underline{\Phi}(\pi) \tag{1.5-40}$$

The eigenvalues of $\underline{C}$ are given by

$$\begin{vmatrix} \phi_{11}(\pi) - \lambda & \phi_{12}(\pi) \\ \phi_{21}(\pi) & \phi_{22}(\pi) - \lambda \end{vmatrix} = 0 \tag{1.5-41a}$$

or $\qquad \lambda^2 - [\phi_{11}(\pi) + \phi_{22}(\pi)]\lambda + \det\underline{\Phi} = 0 \tag{1.5-41b}$

From Equations (1.5-24b, 41b) and noting that $\operatorname{tr}\underline{A}$ is zero [Equation (1.5-39)], we deduce that

$$\lambda_1\lambda_2 = 1, \qquad \lambda_1 + \lambda_2 = \phi_{11}(\pi) + \phi_{22}(\pi) \tag{1.5-42a,b}$$

The condition for Equation (1.5-36) to have a periodic solution with period π is either λ_1 or λ_2 is one. If λ_1 is one, it follows from Equation (1.5-42a) that λ_2 must also be one. From Equation

(1.5-42b), we deduce that the condition on $\underline{\underline{\Phi}}(\pi)$ for the existence of a periodic solution with period π is

$$\text{tr}\,\underline{\underline{\Phi}}(\pi) = 2 \tag{1.5-43}$$

Equation (1.5-36) has a periodic solution with period 2π if

$$\lambda_1 = \lambda_2 = -1, \qquad \text{tr}\,\underline{\underline{\Phi}}(\pi) = -2 \tag{1.5-44a,b,c}$$

We can verify the results we have deduced by considering $p(t)$ to be a constant which is periodic of arbitrary period. We consider two values of $p(t)$.

(a)   Let $p(t) = 1$. Two fundamental solutions are

$$\underline{\phi}_1 = \begin{bmatrix} \cos t \\ -\sin t \end{bmatrix}, \qquad \underline{\phi}_2 = \begin{bmatrix} \sin t \\ \cos t \end{bmatrix} \tag{1.5-45a,b}$$

$\underline{\phi}_1$ and $\underline{\phi}_2$ generate

$$\underline{\underline{\Phi}}(t) = \begin{bmatrix} \cos t & \sin t \\ -\sin t & \cos t \end{bmatrix} \tag{1.5-46a}$$

$$\underline{\underline{\Phi}}(0) = \underline{\underline{I}}, \qquad \underline{\underline{\Phi}}(\pi) = \begin{bmatrix} -1 & 0 \\ 0 & -1 \end{bmatrix} \tag{1.5-46b,c}$$

Since $\text{tr}\,\underline{\underline{\Phi}}(\pi)$ is -2, the solution is periodic with period 2π as given by Equation (1.5-45a, b).

(b)   Let $p(t) = 4$. Two fundamental solutions can be written as

$$\underline{\phi}_1 = \begin{bmatrix} \cos 2t \\ -2\sin 2t \end{bmatrix}, \qquad \underline{\phi}_2 = \begin{bmatrix} (1/2)\sin 2t \\ \cos 2t \end{bmatrix} \tag{1.5-47a,b}$$

From Equations (1.5-47a, b), we obtain

$$\underline{\underline{\Phi}}(t) = \begin{bmatrix} \cos 2t & \dfrac{1}{2}\sin t \\ -2\sin 2t & \cos 2t \end{bmatrix} \tag{1.5-48a}$$

$$\underline{\Phi}\,(0) \,=\, \underline{I}\,, \qquad \underline{\Phi}\,(\pi) \,=\, \begin{bmatrix} 1 & 0 \\ 0 & 1 \end{bmatrix} \qquad\qquad (1.5\text{-}48b,c)$$

In this case, $\mathrm{tr}\,\underline{\Phi}\,(\pi)$ is 2 and the solution is periodic of period π as given by Equation (1.5-47a, b).

●

In this chapter, we have developed qualitative methods of solving differential equations. We have not computed the numerical values of the solutions. In the next chapter, we shall consider approximate methods of obtaining numerical values of the solutions of differential equations. Qualitative and quantitative methods are complementary to each other.

PROBLEMS

1.   Show that the solution of

$$\frac{dx_1}{dt} = -x_1\,, \qquad \frac{dx_2}{dt} = -2x_2$$

satisfying the initial conditions

$$x_1(0) \,=\, 2\,, \qquad x_2(0) \,=\, 4$$

is

$$x_1(t) \,=\, 2e^{-t}\,, \qquad x_2(t) \,=\, 4e^{-2t}$$

Deduce and sketch the trajectory.

Use the chain rule to show that another solution of the differential equations is $[x_1(t+a),\, x_2(t+a)]$, where a is a constant.

Does the trajectory that represent the solution $[x_1(t),\, x_2(t)]$ also represent the solution $[x_1(t+a),\, x_2(t+a)]$?

2.   The equation for the current i in a circuit consisting of a voltage source, a capacitor, an inductor, and a resistor connected in series is [Close and Frederick (1978)]

$$L\frac{d^2 i}{dt^2} + R\frac{di}{dt} + \frac{i}{C} = \frac{de}{dt}$$

where L is the inductance, R is the resistance, C is the capacitance, and e is the voltage source.

Write the differential equation as a system of two first order equations. Is the system autonomous?

Consider the case e = 0. Find the equilibrium point and examine its stability.

3.   The equation of motion of a particle moving under a central attractive force obeying the inverse square law is

$$\frac{d^2u}{d\theta^2} + u = \frac{\gamma}{h^2}$$

where $u = 1/r$, r, and θ are the polar coordinates of the particle, γ is a constant, and h is the angular momentum of the particle about the origin.

Write the equation as a system of two first order equations. Determine the equilibrium point and examine its stability.

4.   In Example 1.3-3, it is shown that the damped harmonic oscillator has multiple equilibrium points and they are given by Equations (1.3-20a, b). Examine the stability of the following equilibrium points:

(i) $x_1 = 2\pi$, $x_2 = 0$ (ii) $x_1 = -\pi$, $x_2 = 0$

5.   Determine the equilibrium point of the following linear systems:

(i) $\dot{x}_1 = x_1 + x_2 - 2$, $\dot{x}_2 = x_1 - x_2$

(ii) $\dot{x}_1 = -(x_1 + x_2 + 1)$, $\dot{x}_2 = 2x_1 - x_2 + 5$

(iii) $\dot{x}_1 = 3x_1 + x_2$, $\dot{x}_2 = -x_1 + x_2$

Describe the type and the stability of each of the equilibrium points.

6.   Determine all the equilibrium points of the following system of equations

$$\frac{dx_1}{dt} = x_1^2 + x_2^2 - 1$$

$$\frac{dx_2}{dt} = 2x_1 x_2$$

Examine the stability of all the equilibrium points.

7.   The spread of an epidemy can be modeled by making the following assumptions. The population is divided into three groups: (a) those who are infected with the disease, (b) those who are susceptible to catch the disease, and (c) those who are immune to the disease. Let $I(t)$ and $S(t)$ be the number of people in groups (a) and (b) respectively. The rate of increase of I (or decrease of S) is proportional to IS. A vaccination program is implemented and as a result S decreases at a rate proportional to S. Due to migration and loss of immunity, S increases at a constant rate μ. These assumptions lead to the following equations for S and I

$$\frac{dS}{dt} = -\alpha SI - \lambda S + \mu$$

$$\frac{dI}{dt} = \alpha SI - \gamma I$$

Determine all the equilibrium points and examine their stability.

8.   The predator-prey model in Example 1.3-4 is modified by writing Equations (1.3-44, 27b) as

$$\frac{dN_1}{dt} = a_1 N_1 (1 - N_1/K) - g(N_1) N_2$$

$$\frac{dN_2}{dt} = b_2 g(N_1) N_2 - a_2 N_2$$

$$g(N_1) = aN_1/(b + N_1)$$

Note that as $N_1 \longrightarrow \infty$, $g(N_1) \longrightarrow a$. Non-dimensionalize the equations, determine the equilibrium points, and discuss their stability. Draw the nullclines and sketch the phase portrait.

9.   Show that the following systems have no periodic solutions:

(i) $\dot{x}_1 = x_1^3 + x_2$, $\dot{x}_2 = x_1 + x_2 + x_2^3$

(ii) $\dot{x}_1 = x_1(x_2 - 1)$, $\dot{x}_2 = x_1 + x_2 - x_2^2/2$

(iii) $\dot{x}_1 = x_2$, $\dot{x}_2 = x_2(1 + x_1^2) + x_1^3$

(iv) $\ddot{x} + b(x^2 + 1)\dot{x} + cx = 0$

10.   Transform the following system of equations into polar form

$$\dot{x}_1 = 4x_1 + 4x_2 - x_1(x_1^2 + x_2^2)$$

$$\dot{x}_2 = -4x_1 + 4x_2 - x_2(x_1^2 + x_2^2)$$

Show that the system has a periodic solution. Determine that solution.

11. Show that Mathieu's equation [Equations (1.5-36, 38)] can be written as

$$
\begin{bmatrix} \dot{x}_1 \\ \dot{x}_2 \end{bmatrix} = \begin{bmatrix} 0 & 1 \\ -\delta - \varepsilon \cos 2t & 0 \end{bmatrix} \begin{bmatrix} x_1 \\ x_2 \end{bmatrix}
$$

Deduce that the characteristic multipliers λ_1 and λ_2 are the roots of

$$\lambda^2 - \phi(\delta, \varepsilon)\lambda + 1 = 0$$

where $\phi(\delta, \varepsilon)$ is not known explicitly.

Show that if

(i) $\phi > 2$, either λ_1 or λ_2 is greater than unity and one of the solutions is unbounded.

(ii) $|\phi| = 2$, a periodic solution exists.

(iii) $-2 < \phi < 2$, λ_1 and λ_2 are complex conjugates. What can be deduced?

(iv) $\phi < -2$, λ_1 and λ_2 are real and negative. Are the solutions bounded?

CHAPTER 2

PERTURBATION METHODS

2.1 REGULAR AND SINGULAR PERTURBATION

Many of the differential equations that are of interest to engineers and scientists cannot be solved exactly. In Chapter 1, we have explained methods of obtaining qualitative properties of the solutions of the equations. In this chapter, we develop approximate methods, involving a small parameter ε, of obtaining quantitative solutions. For example, in fluid mechanics, the Navier-Stokes equation is associated with the Reynolds number Re. For flows at low Reynolds number, we can choose ε to be Re and at high Reynolds number, we choose ε to be $1/\text{Re}$.

We next assume that the solution $x(t; \varepsilon)$ of the differential equation can be expanded as a power series in ε. We differentiate the series term by term and substitute the resulting expression in the differential equations. Comparing powers of ε yields a set of differential equations which, when compared to the original differential equation, might be easier to solve. We illustrate this by considering a non-linear equation which is written as

$$L(x) + \varepsilon N(x) = 0 \tag{2.1-1}$$

where L and N are linear and non-linear operators respectively.

We expand x in a power series of ε as

$$x = x_0(t) + \varepsilon x_1(t) + \dots \tag{2.1-2}$$

Differentiating, substituting the resulting expressions in Equation (2.1-1), and comparing powers of ε yields

$$\varepsilon^0: \quad L(x_0) = 0 \tag{2.1-3a}$$

$$\varepsilon^1: \quad L(x_1) = -N(x_0) \tag{2.1-3b}$$

Equation (2.1-3a) is a linear homogeneous equation and can usually be solved. Equation (2.1-3b) is a linear non-homogeneous equation and can also be solved. By the **perturbation method**, the non-linear equation is simplified to linear equations.

The series solution [Equation (2.1-2)] is often not convergent, unlike the series solution obtained by the method of Frobenius. Usually, the series is an **asymptotic series**. The properties and usefulness of such series can be found in Cesari (1971), Chan Man Fong et al. (1997), and Wilcox (1995).

If the series solution is valid throughout the region of interest, it is a **regular perturbation**, and if it is valid only for a limited range, it is a **singular perturbation**. It is the rule rather than the exception that the parameter series expansion is singular. To render these expressions **uniformly valid** (valid throughout the region), several techniques have been developed. We describe these techniques by solving several examples. We start by considering regular perturbation.

Example 2.1-1. **Regular perturbation**

Solve the one-dimensional steady diffusion equation where the diffusion coefficient (diffusivity) is a function of the concentration. At steady state, the equation governing the diffusion through a plane sheet can be written as (Crank, 1975)

$$\frac{d}{dx}\left(D\,\frac{dc}{dx}\right) = 0 \tag{2.1-4}$$

where D is the diffusion coefficient and c is the concentration.

In many cases, D is an increasing function of c and as a first approximation, we assume that D is a linear function and is written as

$$D = D_0\,(1 + \varepsilon c) \tag{2.1-5}$$

where D_0 and ε are constants.

In a sorption test, the boundary conditions can be written as

$$c\,(0) = c_s, \qquad c\,(\ell) = 0 \tag{2.1-6a,b}$$

where c_s is the saturation concentration and ℓ is the thickness of the sheet.

We assume ε to be small and seek a solution of the form

$$c\,(x) = c_0(x) + \varepsilon c_1(x) + \dots \tag{2.1-7}$$

Combining Equations (2.1-4, 5, 7) and comparing powers of ε yields

$$\frac{d}{dx}\left(D_0\,\frac{dc_0}{dx}\right) = 0 \tag{2.1-8a}$$

$$\frac{d}{dx}\left[D_0\left(c_0\,\frac{dc_0}{dx} + \frac{dc_1}{dx}\right)\right] = 0 \tag{2.1-8b}$$

Equations (2.1-6a,b) become

$$c_0(0) = c_s, \qquad c_0(\ell) = 0 \qquad\qquad\qquad (2.1\text{-}9a,b)$$

$$c_1(0) = 0, \qquad c_1(\ell) = 0 \qquad\qquad\qquad (2.1\text{-}9c,d)$$

Equations (2.1-9c, d) result from assuming that c_s is of order ε^0.

The solution of Equation (2.1-8a) subject to (2.1-9a, b) is

$$c_0 = c_s(1 - x/\ell) \qquad\qquad\qquad (2.1\text{-}10)$$

Substituting c_0 into Equation (2.1-8b) yields

$$\frac{d^2 c_1}{dx^2} = -\left(\frac{c_s}{\ell}\right)^2 \qquad\qquad\qquad (2.1\text{-}11)$$

The solution, subject to Equations (2.1-9c, d), is

$$c_1 = \frac{c_s^2 \, x}{2\ell}\left(1 - \frac{x}{\ell}\right) \qquad\qquad\qquad (2.1\text{-}12)$$

Combining Equations (2.1-7, 10, 12) yields

$$c = c_s\left[(1 - x/\ell)(1 + \varepsilon c_s x/2\ell)\right] + \dots \qquad\qquad\qquad (2.1\text{-}13)$$

In this example, we can obtain an exact solution. Integrating Equation (2.1-4) yields

$$D\frac{dc}{dx} = K_0 \qquad\qquad\qquad (2.1\text{-}14)$$

where K_0 is a constant.

Substituting Equation (2.1-5) into Equation (2.1-14), integrating the resulting expression, and imposing the boundary conditions, we obtain

$$c + \varepsilon c^2/2 = c_s(1 - x/\ell)(1 + \varepsilon c_s/2) \qquad\qquad\qquad (2.1\text{-}15)$$

We deduce that c is given by

$$c = \frac{1}{\varepsilon}\left[-1 \pm \sqrt{1 + 2\varepsilon c_s(1 - x/\ell)(1 + \varepsilon c_s/2)}\,\right] \qquad\qquad\qquad (2.1\text{-}16)$$

The terms inside the square root can be expanded as

$$[1 + 2\varepsilon c_s (1 - x/\ell)(1 + \varepsilon c_s/2)]^{1/2}$$

$$= 1 + \varepsilon c_s (1 - x/\ell)(1 + \varepsilon c_s/2) - (1/2)[\varepsilon c_s (1 - x/\ell)(1 + \varepsilon c_s/2)]^2 + ... \qquad (2.1\text{-}17a)$$

$$= 1 + \varepsilon c_s (1 - x/\ell) + \varepsilon^2 c_s^2 (1 - x/\ell)(x/2\ell) + ... \qquad (2.1\text{-}17b)$$

Equation (2.1-16) implies the existence of two solutions and on physical ground, we have to choose the positive sign (c is positive). On substituting Equation (2.1-17b) into Equation (2.1-16), we obtain Equation (2.1-13). Note that if in Equation (2.1-16), the negative sign is chosen, c will be singular, that is to say, $|c| \longrightarrow \infty$ as $\varepsilon \longrightarrow 0$ and c cannot be expanded as a power series in ε. We may conclude that the straightforward expansion [Equation (2.1-7)] yields the regular solution.

In this example, the solution is valid for all values of x in the interval $0 \le x \le \ell$. In the next example, we consider the case where the solution is valid for only a certain range of x.

●

Example 2.1-2. **Secular terms**

Solve the following equation by the perturbation method

$$\frac{d^2 x}{dt^2} + (1 + \varepsilon) x = 0 \qquad (2.1\text{-}18a)$$

subject to

$$x(0) = 1, \qquad \dot{x}(0) = 0 \qquad (2.1\text{-}18b,c)$$

where ε is a constant.

We expand x as

$$x = x_0 + \varepsilon x_1 + ... \qquad (2.1\text{-}19)$$

Substituting Equation (2.1-19) into Equation (2.1-18a) and comparing powers of ε yields

$$\frac{d^2 x_0}{dt^2} + x_0 = 0 \qquad (2.1\text{-}20a)$$

$$\frac{d^2 x_1}{dt^2} + x_1 = -x_0 \qquad (2.1\text{-}20b)$$

Equations (2.1-18b, c) become

$$x_0(0) = 1, \qquad \dot{x}_0(0) = 0 \qquad\qquad\qquad (2.1\text{-}21a,b)$$

$$x_1(0) = 0, \qquad \dot{x}_1(0) = 0 \qquad\qquad\qquad (2.1\text{-}21c,d)$$

The solutions of Equations (2.1-20a, b, 21a–d) are

$$x_0 = \cos t \qquad\qquad\qquad (2.1\text{-}22a)$$

$$x_1 = -(t/2)\sin t \qquad\qquad\qquad (2.1\text{-}22b)$$

Combining Equations (2.1-19, 22a, b) yields

$$x = \cos t - \varepsilon(t/2)\sin t + ... \qquad\qquad\qquad (2.1\text{-}23)$$

From Equation (2.1-23), we deduce that as $t \longrightarrow \infty$, x oscillates between $\pm\infty$. However, Equation (2.1-18a) is the simple harmonic equation and its solution subject to Equations (2.1-18b, c) is

$$x = \cos\left[(\sqrt{1+\varepsilon})\,t\right] \qquad\qquad\qquad (2.1\text{-}24)$$

The exact solution is periodic and x is bounded for all values of t. The first term on the right side of Equation (2.1-23) is periodic (bounded), the second term is not bounded and is the **secular term**. This term ceases to be a valid approximation when t is of the order of $(1/\varepsilon)$.

We can recover the approximate solution from the exact solution by expanding $\cos\left[(\sqrt{1+\varepsilon})\,t\right]$ in the following way

$$\cos\left[(\sqrt{1+\varepsilon})\,t\right] = \cos(1 + \varepsilon/2 + ...)t \qquad\qquad (2.1\text{-}25a)$$

$$= \cos t \cos(\varepsilon t/2) - \sin t \sin(\varepsilon t/2) + ... \qquad (2.1\text{-}25b)$$

$$= \cos t - (\varepsilon t/2)\sin t + ... \qquad\qquad (2.1\text{-}25c)$$

Equation (2.1-25b) is obtained by using the trigonometric relations for $\cos(A+B)$ and Equation (2.1-25c) results from the expansion of $\cos(\varepsilon t/2)$ and $\sin(\varepsilon t/2)$ in powers of ε.

The secular term appears when $\sin(\varepsilon t/2)$ is approximated by $(\varepsilon t/2)$. The expansion of $\sin(\varepsilon t/2)$ in powers of $(\varepsilon t/2)$ is known to be convergent. However, when $t = O(1/\varepsilon)$, the convergence is very slow and for practical purposes, the series is not useful. Adding higher order terms does not solve the problem. For $t \geq O(1/\varepsilon)$, $(\varepsilon t/2)$ is of $O(1)$ and each succeeding term is not smaller than the previous term. The magnitude of the term neglected can be greater than that of the terms which are retained and the solution cannot be approximated by a finite number of terms. To extend the solution to all values of t, we need to develop new techniques and these will be discussed in subsequent sections.

●

In Example 2.1-2, the non-uniformity manifests itself through the secular term and the approximate solution becomes invalid because we wish to extend the solution for large t $[t > O(1/\varepsilon)]$, that is to say, the domain is allowed to be infinite. Another case that can give rise to non-uniformity is when the coefficient of the highest derivative in the differential equation is ε. On seeking a solution as a power series in ε, the zeroth order equation is an order lower than the original differential equation and it is not possible to satisfy all the boundary and initial conditions, as shown in the next example.

Example 2.1-3. **Singular perturbation**

Use perturbation methods to solve the Michaelis-Menten (1913) enzymatic reactions.

The Michaelis-Menten reaction involves a substrate S reacting with an enzyme E to form an intermediate complex SE which in turn is broken down to a product P and the enzyme. The reactions can be represented by

$$S + E \underset{k_{-1}}{\overset{k_1}{\rightleftharpoons}} SE$$

$$SE \overset{k_2}{\longrightarrow} P + E$$

The kinetic equations for the reactions are

$$\frac{d[S]}{dt} = -k_1[E][S] + k_{-1}[SE] \tag{2.1-26a}$$

$$\frac{d[E]}{dt} = -k_1[E][S] + (k_{-1} + k_2)[SE] \tag{2.1-26b}$$

$$\frac{d[SE]}{dt} = k_1[E][S] - (k_{-1} + k_2)[SE] \tag{2.1-26c}$$

$$\frac{d[P]}{dt} = k_2[SE] \tag{2.1-26d}$$

The initial conditions are

$$[S(0)] = [S_0], \qquad [E(0)] = [E_0] \tag{2.1-27a,b}$$

$$[SE(0)] = [P(0)] = 0 \tag{2.1-27c,d}$$

The two conservation equations [adding Equations (2.1-26b, c) and (26a, c, d)] are

$$[E] + [SE] = \text{constant} = [E_0] \qquad (2.1\text{-}28a)$$

$$[S] + [SE] + [P] = \text{constant} = [S_0] \qquad (2.1\text{-}28b)$$

Equation (2.1-28a) implies that

$$[E] = [E_0] - [SE] \qquad (2.1\text{-}29)$$

Note that once $[SE]$ is known, $[P]$ and $[E]$ can be determined from Equations (2.1-26d, 29). We need to solve Equations (2.1-26a, c). Substituting Equation (2.1-29) into Equations (2.1-26a, c) yields

$$\frac{d[S]}{dt} = -k_1[E_0][S] + \{k_1[S] + k_{-1}\}[SE] \qquad (2.1\text{-}30a)$$

$$\frac{d[SE]}{dt} = k_1\{[E_0] - [SE]\}[S] - (k_{-1} + k_2)[SE] \qquad (2.1\text{-}30b)$$

We introduce the following dimensionless quantities

$$\tau = k_1[E_0]t, \qquad u = [S]/[S_0], \qquad v = [SE]/[E_0] \qquad (2.1\text{-}31a,b,c)$$

$$\lambda_1 = k_{-1}/k_1[S_0], \qquad \lambda_2 = (k_1 + k_2)/k_1[S_0], \qquad \varepsilon = [E_0]/[S_0] \qquad (2.1\text{-}31d,e,f)$$

Combining Equations (2.1-30a – 31f) yields

$$\frac{du}{d\tau} = -u + (u + \lambda_1)v \qquad (2.1\text{-}32a)$$

$$\varepsilon\frac{dv}{d\tau} = u - (u + \lambda_2)v \qquad (2.1\text{-}32b)$$

The appropriate initial conditions are

$$u(0) = 1, \qquad v(0) = 0 \qquad (2.1\text{-}33a,b)$$

The quantity ε is usually in the range 10^{-2} to 10^{-7} (Murray 1993) and we seek a solution of the form

$$u = u_0 + \varepsilon u_1 + \ldots \qquad (2.1\text{-}34a)$$

$$v = v_0 + \varepsilon v_1 + \ldots \qquad (2.1\text{-}34b)$$

Differentiating u and v, substituting the resulting expressions into Equations (2.1-32a, b), and comparing powers of ε yields the following zeroth order equations

$$\frac{du_0}{d\tau} = -u_0 + (u_0 + \lambda_1)v_0 \qquad (2.1\text{-}35a)$$

$$0 = u_0 - (u_0 + \lambda_2) v_0 \qquad (2.1\text{-}35b)$$

The initial conditions are

$$u_0(0) = 1, \qquad v_0(0) = 0 \qquad (2.1\text{-}36a,b)$$

Note that since the coefficient of the derivative in Equation (2.1-32b) is ε, its zeroth order equation is an algebraic equation [Equation (2.1-35b)]. Its solution does not introduce an arbitrary constant and is

$$v_0 = u_0 / (u_0 + \lambda_2) \qquad (2.1\text{-}37)$$

Substituting Equation (2.1-37) into Equation (2.1-35a) yields

$$(u_0 + \lambda_2) \frac{du_0}{d\tau} = (\lambda_1 - \lambda_2) u_0 \qquad (2.1\text{-}38)$$

The solution is

$$(u_0 + \lambda_2 \ell n\, u_0) = (\lambda_1 - \lambda_2)\, \tau + C_0 \qquad (2.1\text{-}39)$$

where C_0 is a constant.

Imposing the initial condition, Equation (2.1-39) becomes

$$(u_0 + \lambda_2 \ell n\, u_0) = 1 + (\lambda_1 - \lambda_2)\, \tau \qquad (2.1\text{-}40)$$

The function u_0 satisfies the initial condition but v_0 does not. From Equations (2.1-36a, 37), we deduce that

$$v_0(0) = 1 / (1 + \lambda_2) \neq 0 \qquad (2.1\text{-}41a,b)$$

This implies that the solution is not valid near $\tau = 0$. There is a boundary (initial) layer in the neighborhood of the origin $(\tau = 0)$ where v_0 changes rapidly and $\varepsilon\,(dv_0/d\tau)$ is of $O(1)$ and not of $O(\varepsilon)$. We need to introduce another time scale such that the zeroth order equation for v [Equation (2.1-32b)] is a differential equation and the integration constant is chosen so as to satisfy the initial condition. The solution we obtain by straightforward expansion is the pseudo-steady state solution and is not valid at the beginning and at the end of the reaction. A more detailed discussion on the validity of the present solution is given in Roberts (1977).

●

The perturbation method can also be used to solve partial differential equations and this is illustrated in the next two examples.

Example 2.1-4. **Partial differential equation**

Solve the following boundary-value problem by the method of perturbation

$$\frac{\partial^2 u}{\partial r^2} + \frac{1}{r}\frac{\partial u}{\partial r} + \frac{1}{r^2}\frac{\partial^2 u}{\partial \theta^2} + \varepsilon u^2 = 0, \quad 0 \le r \le 1, \quad 0 \le \theta \le 2\pi \tag{2.1-42a}$$

$$u(1, \theta) = \cos\theta \tag{2.1-42b}$$

u is finite at the origin.

We assume that ε is small and expand u as

$$u(r, \theta) = u_0(r, \theta) + \varepsilon u_1(r, \theta) + \dots \tag{2.1-43}$$

Differentiating, substituting the resulting expressions into Equation (2.1-42a), and comparing powers of ε yields

$$\varepsilon^0: \quad \frac{\partial^2 u_0}{\partial r^2} + \frac{1}{r}\frac{\partial u_0}{\partial r} + \frac{1}{r^2}\frac{\partial^2 u_0}{\partial \theta^2} = 0 \tag{2.1-44a}$$

$$\varepsilon: \quad \frac{\partial^2 u_1}{\partial r^2} + \frac{1}{r}\frac{\partial u_1}{\partial r} + \frac{1}{r^2}\frac{\partial^2 u_1}{\partial \theta^2} = -u_0^2 \tag{2.1-44b}$$

Equation (2.1-42b) becomes

$$u_0(1, \theta) = \cos\theta, \qquad u_1(1, \theta) = 0 \tag{2.1-45a,b}$$

The functions u_0 and u_1 are finite at the origin.

Equation (2.1-44a) can be solved by the method of separation of variables. We write u_0 as

$$u_0(r, \theta) = R(r)\, G(\theta) \tag{2.1-46}$$

Differentiating, substituting the resulting expressions into Equation (2.1-44a), and separating the variables yields

$$r^2 R'' + r R' - n^2 R = 0 \tag{2.1-47a}$$

$$G'' + n^2 G = 0 \tag{2.1-47b}$$

where n^2 is a constant (constant of separation) and prime (') denotes differentiation with respect to the argument.

Equation (2.1-47a) is Euler's equation and we seek a solution of the form

$$R = r^s \qquad (2.1\text{-}48)$$

Combining Equations (2.1-47a, 48) yields a quadratic equation which can be written as

$$s^2 - n^2 = 0 \qquad (2.1\text{-}49)$$

The two solutions are

$$s = \pm n \qquad (2.1\text{-}50)$$

From Equations (2.1-48, 50), we deduce that R can be written as

$$R = C_0 r^n + C_1 r^{-n} \qquad (2.1\text{-}51)$$

where C_0 and C_1 are constants.

The condition that u_0 is finite at the origin implies that

$$C_1 = 0 \qquad (2.1\text{-}52)$$

The solution of Equation (2.1-47b) is

$$G = D_0 \cos n\theta + D_1 \sin n\theta \qquad (2.1\text{-}53)$$

where D_0 and D_1 are constants.

Substituting Equations (2.1-51 – 53) into Equation (2.1-46) yields

$$u_0 = C_0 r^n (D_0 \cos n\theta + D_1 \sin n\theta) \qquad (2.1\text{-}54)$$

The constants n, $C_0 D_0$, and $C_0 D_1$ are determined by imposing the boundary condition [Equation (2.1-45a)] and u_0 is found to be

$$u_0 = r \cos \theta \qquad (2.1\text{-}55)$$

Combining Equations (2.1-44b, 55) yields

$$\frac{\partial^2 u_1}{\partial r^2} + \frac{1}{r} \frac{\partial u_1}{\partial r} + \frac{1}{r^2} \frac{\partial^2 u_1}{\partial \theta^2} = -r^2 \cos^2 \theta = -\frac{r^2}{2} (1 + \cos 2\theta) \qquad (2.1\text{-}56\text{a,b})$$

The right side of Equation (2.1-56b) suggests that we seek a solution of the form

$$u_1 = f(r) + g(r) \cos 2\theta \qquad (2.1-57)$$

Differentiating, substituting the resulting expressions into Equation (2.1-56b), and comparing terms with and without $\cos 2\theta$ results in two equations which can be written as

$$r f'' + f' = -r^3/2 \qquad (2.1-58a)$$

$$r^2 g'' + r g' - 4g = -r^4/2 \qquad (2.1-58b)$$

The solution of Equation (2.1-58a) which is finite at the origin is

$$f = A_0 - r^4/32 \qquad (2.1-59)$$

where A_0 is a constant.

Equation (2.1-58b) is a non-homogeneous Euler's equation and the particular integral can be obtained by the method of variation of parameters (Chan Man Fong et al., 1997). The solution which is finite at the origin is

$$g = A_1 r^2 - r^4/24 \qquad (2.1-60)$$

Combining Equations (2.1-57, 59, 60) yields

$$u_1 = A_0 - r^4/32 + (A_1 r^2 - r^4/24) \cos 2\theta \qquad (2.1-61)$$

From Equations (2.1-45b, 61), we deduce that

$$A_0 = 1/32, \qquad A_1 = 1/24 \qquad (2.1-62a,b)$$

The solution u to order of ε is obtained from Equations (2.1-43, 55, 57, 59 to 62b) and can be written as

$$u = r \cos \theta + \varepsilon [(1 - r^4)/32 + r^2 \cos 2\theta \, (1 - r^2)/24] + ... \qquad (2.1-63)$$

The solution is valid throughout the region and the perturbation is regular.

●

Example 2.1-5. **Perturbed boundaries**

Use the method of perturbation to solve

$$\frac{\partial^2 u}{\partial x^2} + \frac{\partial^2 u}{\partial y^2} = 0, \quad 0 < x < \pi, \quad y > \varepsilon x \qquad (2.1\text{-}64a)$$

$$u(0, y) = u(\pi, y) = 0 \qquad (2.1\text{-}64b,c)$$

$$u(x, \varepsilon x) = \sin x, \quad u \longrightarrow 0 \ \text{ as } \ y \longrightarrow \infty \qquad (2.1\text{-}64d,e)$$

In this example, the region of interest is not rectangular, it is trapezoidal as shown in Figure 2.1-1. We consider the boundary OA to be a perturbation of the x-axis.

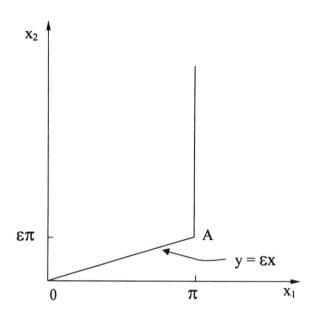

FIGURE 2.1-1   Perturbed boundaries

We seek a solution of the form

$$u = u_0 + \varepsilon u_1 + \dots \qquad (2.1\text{-}65)$$

By the usual procedure, we obtain

$$\varepsilon^0: \quad \frac{\partial^2 u_0}{\partial x^2} + \frac{\partial^2 u_0}{\partial y^2} = 0 \qquad (2.1\text{-}66a)$$

$$\varepsilon: \quad \frac{\partial^2 u_1}{\partial x^2} + \frac{\partial^2 u_1}{\partial y^2} = 0 \qquad (2.1\text{-}66b)$$

Equations (2.1-64b, c, e) become

$$u_0(0, y) = u_1(0, y) = u_0(\pi, y) = u_1(\pi, y) = 0 \qquad (2.1\text{-}67\text{a-d})$$

$$u_0 \longrightarrow 0, \qquad u_1 \longrightarrow 0 \qquad \text{as } y \longrightarrow \infty \qquad (2.1\text{-}67\text{e,f})$$

The boundary condition given by Equation (2.1-64d) is rewritten by expanding $u(x, \varepsilon x)$ in a Taylor series about $y = 0$ and Equation (2.1-64d) becomes

$$u_0(x, 0) + \varepsilon x \left. \frac{\partial u_0}{\partial y} \right|_{y=0} + \dots + \varepsilon u_1(x, 0) + \dots = \sin x \qquad (2.1\text{-}68)$$

From Equation (2.1-68), we deduce that

$$u_0(x, 0) = \sin x, \qquad u_1(x, 0) = -x \left. \frac{\partial u_0}{\partial y} \right|_{y=0} \qquad (2.1\text{-}69\text{a,b})$$

Equation (2.1-66a) can be solved by the method of separation of variables. The solution that satisfies Equations (2.1-67a, c, e, 69a) is

$$u_0 = e^{-y} \sin x \qquad (2.1\text{-}70)$$

The general solution of Equation (2.1-66b) subject to Equations (2.1-67b, d, f) is

$$u_1 = \sum_{n=1}^{\infty} b_n e^{-ny} \sin nx \qquad (2.1\text{-}71)$$

where b_n $(n = 1, 2, \dots)$ are constants.

Imposing the boundary condition given by Equation (2.1-69b) yields

$$\sum_{n=1}^{\infty} b_n \sin nx = x \sin x \qquad (2.1\text{-}72)$$

The Fourier coefficients b_n are given by

$$b_n = \frac{2}{\pi} \int_0^{\pi} x \sin x \sin nx \, dx \qquad (2.1\text{-}73)$$

Integrating yields

$$b_n = \begin{cases} \pi/2, & n = 1 \\[4mm] \dfrac{2n}{(n^2 - 1)^2} [(-1)^{n+1} - 1], & n \neq 1 \end{cases} \qquad (2.1\text{-}74a,b)$$

We note from Equation (2.1-74b) that b_n is zero if n $(\neq 1)$ is odd.

Combining Equations (2.1-70, 71, 74a, b) yields

$$u = e^{-y} \sin x + \varepsilon \left[(\pi/2) e^{-y} \sin x - 4 \sum_{s=1}^{\infty} \frac{s\, e^{-2sy}}{(4s^2 - 1)^2} \sin 2sx \right] + ... \qquad (2.1\text{-}75)$$

The solution is valid throughout the region.

●

In Examples 2.1-2 and 3, we have shown that although ε is small, the method of perturbation yields solutions which are not valid throughout the region. A similar situation exist in the case of partial differential equations. Examples of singular perturbation involving partial differential equations will be considered in subsequent sections.

We introduce the method of multiple scales in the next section.

2.2 METHOD OF MULTIPLE SCALES

In Examples 2.1-2 and 3, we have discussed various sources of non-uniformity of solutions of differential equations. We have proposed that we need to introduce additional time scales so as to obtain uniformly valid solutions. Suppose that the independent function $x(t)$ is treated as a function of two independent variables T_0 and T_1, that is, we have introduced two time scales T_0 and T_1. The ordinary differential equations now become partial differential equations and on integrating these equations, we introduce arbitrary functions. These functions are determined by requiring that the secular terms should vanish. The solution so obtained is uniformly valid. We illustrate this method, **method of multiple scales**, by solving a few problems.

Example 2.2-1. **Regularization of solutions of previous examples**

Use the method of multiple scales to obtain a uniformly valid solution of Equations (2.1-18a–c, 32a–33b).

In Example 2.1-2, we observed that the solution obtained by straightforward expansion is not valid when $t = O(1/\varepsilon)$. This suggests that we need to introduce a time scale that contracts the time so that large values of t appear to be of order unity in this new time scale. For simplicity, we use two time scales T_0 and T_1 defined by

$$T_0 = t, \qquad T_1 = \varepsilon t \qquad\qquad (2.2\text{-}1a,b)$$

Although there is a relationship between T_0 and T_1 ($T_1 = \varepsilon T_0$), we regard T_0 and T_1 as independent variables. The dependent variable $x(t)$ in Equation (2.1-18a) is now treated as $X(T_0, T_1)$, a function of two independent variables. The usual chain rule yields

$$\frac{dx}{dt} = \frac{\partial X}{\partial T_0} \frac{dT_0}{dt} + \frac{\partial X}{\partial T_1} \frac{dT_1}{dt} \qquad\qquad (2.2\text{-}2a)$$

$$= \frac{\partial X}{\partial T_0} + \varepsilon \frac{\partial X}{\partial T_1} \qquad\qquad (2.2\text{-}2b)$$

Similarly, we obtain

$$\frac{d^2x}{dt^2} = \frac{\partial^2 X}{\partial T_0^2} + 2\varepsilon \frac{\partial^2 X}{\partial T_0 \partial T_1} + \varepsilon^2 \frac{\partial^2 X}{\partial T_1^2} \qquad\qquad (2.2\text{-}3)$$

We expand X in powers of ε and write

$$X = X_0(T_0, T_1) + \varepsilon X_1(T_0, T_1) + \ldots \qquad\qquad (2.2\text{-}4)$$

Combining Equations (2.1-18a, 2-3, 4) and comparing powers of ε yields

$$\varepsilon^0: \qquad \frac{\partial^2 X_0}{\partial T_0^2} + X_0 = 0 \qquad\qquad (2.2\text{-}5a)$$

$$\varepsilon: \qquad \frac{\partial^2 X_1}{\partial T_0^2} + X_1 = -\left(X_0 + 2\frac{\partial^2 X_0}{\partial T_0 \partial T_1} \right) \qquad\qquad (2.2\text{-}5b)$$

The solution of Equation (2.2-5a) can be written as

$$X_0 = A_0(T_1) e^{iT_0} + \overline{A}_0(T_1) e^{-iT_0} \qquad\qquad (2.2\text{-}6a)$$

$$= A_0(T_1) e^{iT_0} + \text{c.c.} \qquad\qquad (2.2\text{-}6b)$$

where A_0 is an arbitrary complex function of T_1, $\overline{A}_0$ is the complex conjugate of A_0 (since we require X_0 to be real), and c.c. denotes complex conjugate. Differentiating X_0 and substituting the resulting expression into Equation (2.2-5b) yields

$$\frac{\partial^2 X_1}{\partial T_0^2} + X_1 = -\left(A_0 e^{i T_0} + 2i \frac{dA_0}{dT_1} e^{i T_0} + \text{c.c.} \right) \tag{2.2-7}$$

The non-homogeneous term gives rise to secular terms and to eliminate them we require that

$$2i \frac{dA_0}{dT_1} + A_0 = 0 \tag{2.2-8}$$

The solution is

$$A_0 = C_0 e^{i T_1/2} \tag{2.2-9}$$

where C_0 is an arbitrary complex number (constant).

Combining Equations (2.2-6a, 9) yields

$$X_0 = C_0 \exp\left[i\,(T_0 + T_1/2) \right] + \overline{C}_0 \exp\left[-i\,(T_0 + T_1/2) \right] \tag{2.2-10a}$$

$$= 2C_r \cos\,(T_0 + T_1/2) - 2C_i \sin\,(T_0 + T_1/2) \tag{2.2-10b}$$

where C_r and C_i are the real and imaginary parts of C_0 respectively.

The initial conditions [Equations (2.1-21a, b)] become

$$X_0\,(0, 0) = 1 , \qquad \left. \frac{\partial X_0}{\partial T_0} \right|_{0,0} = 0 \tag{2.2-11a,b}$$

Imposing the initial conditions leads to

$$X_0 = \cos\,(T_0 + T_1/2) \tag{2.2-12a}$$

or $\quad x_0 = \cos\,(t + \varepsilon t/2) \tag{2.2-12b}$

The solution is periodic and by comparing the present solution with Equations (2.1-24, 25a), we conclude that the solution is valid to order $t = O\,(1/\varepsilon)$. Higher order approximations can be generated in the same way.

In Example 2.1-3, we have seen that the regular perturbation yields a solution that is not valid in the neighborhood of $\tau = 0$. There is a boundary layer of thickness $O\,(\varepsilon)$ near $\tau = 0$ and we introduce a new time scale that magnifies the layer to $O\,(1)$. The time scales we need to consider are

$$T_0 = \tau, \qquad T_1 = \tau/\varepsilon \tag{2.2-13a,b}$$

The functions $u(\tau)$ and $v(\tau)$ are now written as $U(T_0, T_1)$ and $V(T_0, T_1)$ respectively. The operator $d/d\tau$ is given by

$$\frac{d}{d\tau} = \frac{\partial}{\partial T_0} + \frac{1}{\varepsilon} \frac{\partial}{\partial T_1} \tag{2.2-14}$$

The functions U and V are expanded as

$$U = U_0(T_0, T_1) + \varepsilon U_1 + \dots \tag{2.2-15a}$$

$$V = V_0(T_0, T_1) + \varepsilon V_1 + \dots \tag{2.2-15b}$$

Combining Equations (2.1-32a, b, 2-14, 15a, b) and comparing powers of ε yields the zeroth order equations which can be written as

$$\frac{\partial U_0}{\partial T_1} = 0 \tag{2.2-16a}$$

$$\frac{\partial V_0}{\partial T_1} = U_0 - (\lambda_2 + U_0) V_0 \tag{2.2-16b}$$

The first order equations are

$$\frac{\partial U_1}{\partial T_1} = -\frac{\partial U_0}{\partial T_0} - U_0 + (\lambda_1 + U_0) V_0 \tag{2.2-17a}$$

$$\frac{\partial V_1}{\partial T_1} = -\frac{\partial V_0}{\partial T_0} + (1 - V_0) U_1 - (\lambda_2 + U_0) V_1 \tag{2.2-17b}$$

The solution of Equation (2.2-16a) is

$$U_0 = A_0(T_0) \tag{2.2-18}$$

where A_0 is an arbitrary function of T_0.

Equation (2.2-16b) is a first order equation and its integrating factor is $\exp[(\lambda_2 + A_0) T_1]$. The solution is

$$V_0 = A_0 / (\lambda_2 + A_0) + B_0(T_0) \exp[-(\lambda_2 + A_0) T_1] \tag{2.2-19}$$

where B_0 is an arbitrary function of T_0.

To determine A_0 and B_0, we consider the first order equation. Substituting U_0 and V_0 and their derivatives into Equations (2.2-17a, b) yields

$$\frac{\partial U_1}{\partial T_1} = -\frac{dA_0}{dT_0} + \frac{(\lambda_1 - \lambda_2) A_0}{A_0 + \lambda_2} + B_0 (A_0 + \lambda_1) \exp\left[-(\lambda_2 + A_0) T_1\right] \qquad (2.2\text{-}20a)$$

$$\frac{\partial V_1}{\partial T_1} = \frac{\lambda_2}{(A_0 + \lambda_2)^2} \frac{dA_0}{dT_0} + \left(\frac{dB_0}{dT_0} - B_0 T_1 \frac{dA_0}{dT_0}\right) \exp\left[-(\lambda_2 + A_0) T_1\right]$$

$$+ \left\{\frac{\lambda_2}{\lambda_2 + A_0} - B_0 \exp\left[-(\lambda_2 + A_0) T_1\right]\right\} U_1 - (\lambda_2 + A_0) V_1 \qquad (2.2\text{-}20b)$$

The initial conditions [Equations (2.1-33a, b)] are now

$$U_0 (0, 0) = 1 , \qquad V_0 (0, 0) = 0 \qquad (2.2\text{-}21a,b)$$

$$U_1 (0, 0) = 0 , \qquad V_1 (0, 0) = 0 \qquad (2.2\text{-}21c,d)$$

The first two terms on the right side of Equation (2.2-20a) are functions of T_0 only and they generate secular terms. To eliminate these terms, we require

$$\frac{dA_0}{dT_0} = \frac{(\lambda_1 - \lambda_2) A_0}{(A_0 + \lambda_2)} \qquad (2.2\text{-}22a)$$

or $\qquad (1 + \lambda_2 / A_0) \, dA_0 = (\lambda_1 - \lambda_2) \, dT_0 \qquad (2.2\text{-}22b)$

Integrating and imposing Equation (2.2-21a) yields (note $U_0 = A_0$)

$$U_0 + \lambda_2 \, \ell n \, U_0 = (\lambda_1 - \lambda_2) \, T_0 + 1 \qquad (2.2\text{-}23)$$

Equation (2.2-23) is exactly the zeroth solution of u obtained by straightforward expansion [Equation (2.1-40)]. From Equations (2.1-37, 2-19), we deduce that $V_0 \longrightarrow v_0$ in the limit as $T_1 \longrightarrow \infty$ ($\tau \longrightarrow \infty$ or $\varepsilon \longrightarrow 0$).

The exponential change in V_0 takes place in a time τ of $O(\varepsilon)$ and in dimensional time, t is of $O(1/k_1 [S_0])$ which is very short. In many experimental situations, this exponential change is not observed and the quasi-steady state approximation assumed in Example 2.1-3 is valid (Roberts, 1977).

To complete the solution, we have to determine B_0 and this can be achieved by requiring V_1 to be free of secular terms. This method is long and it is easier to obtain V_0 by the method of matched asymptotic expansion which will be explained in Section 4.

In the examples we have considered so far, the small parameter ε occurs in the differential equation and in the boundary conditions. There are many problems of engineering interest where the boundaries are not parallel but the variations are slow. Van Dyke (1987) has given several such examples. In the next example, we consider the flow of a fluid in a channel of slowly varying cross-section.

Example 2.2-2. **Flow in a channel of varying cross-section**

Solve the flow of a Newtonian fluid between two non-parallel planes.

We assume that the flow is two-dimensional, steady, laminar, and at negligible Reynolds number. Under these conditions, the stream function $\psi(x, y)$ satisfies the equation (Bird et al., 1960)

$$\frac{\partial^4 \psi}{\partial x^4} + 2 \frac{\partial^4 \psi}{\partial x^2 \partial y^2} + \frac{\partial^4 \psi}{\partial y^4} = 0 \qquad (2.2\text{-}24)$$

Note that by introducing the stream function $\psi(x, y)$, we have replaced the two dependent velocity components $v_x \ (= \partial \psi / \partial y)$ and $v_y \ (= -\partial \psi / \partial x)$ by a single dependent variable ψ. Further, the equation of continuity is automatically satisfied. The stream function is a constant along a stream line and this implies that ψ is a constant along a fixed boundary. By an appropriate scaling, it can be assumed that ψ is unity along a boundary.

The flow is assumed to be in the x-direction and between two non-parallel walls given by

$$y = \pm h(\varepsilon x) \qquad (2.2\text{-}25)$$

where ε is a small parameter.

Equation (2.2-25) indicates that the boundaries change very slowly with x. From the symmetry of the problem, we need to consider only the region $0 \le y \le h$. The boundary conditions can be written as

$$\psi = 1, \qquad \frac{\partial \psi}{\partial y} = 0 \qquad \text{at} \ \ y = h(\varepsilon x) \qquad (2.2\text{-}26a,b)$$

$$\psi = \frac{\partial^2 \psi}{\partial y^2} = 0 \qquad \text{at} \ \ y = 0 \qquad (2.2\text{-}26c,d)$$

Since the variation in x is slow, we introduce a new scale X defined by

$$X = \varepsilon x \qquad (2.2\text{-}27)$$

The stream function $\psi(x, y)$ is now written as $\Psi(X, y)$ and Equation (2.2-24) becomes

$$\varepsilon^4 \frac{\partial^4 \Psi}{\partial X^4} + 2\varepsilon^2 \frac{\partial^4 \Psi}{\partial X^2 \partial y^2} + \frac{\partial^4 \Psi}{\partial y^4} = 0 \qquad\qquad (2.2\text{-}28)$$

Equations (2.2-26a, b) remain unchanged except for replacing ψ by Ψ and $h(\varepsilon x)$ by $H(X)$.

We seek a solution as a power series in ε^2 and write

$$\Psi = \Psi_0 + \varepsilon^2 \Psi_1 + \dots \qquad\qquad (2.2\text{-}29)$$

The zeroth order equation is

$$\frac{\partial^4 \Psi_0}{\partial y^4} = 0 \qquad\qquad (2.2\text{-}30)$$

The solution is

$$\Psi_0 = (y^3/6)\, A_0(X) + (y^2/2)\, B_0(X) + y\, C_0(X) + D_0(X) \qquad\qquad (2.2\text{-}31a)$$

where $A_0 - D_0$ are arbitrary functions.

Imposing conditions (2.2-26a–d) yields

$$\Psi_0 = \frac{3}{2}\left(\frac{y}{H}\right) - \frac{1}{2}\left(\frac{y}{H}\right)^3 \qquad\qquad (2.2\text{-}31b)$$

The components of the velocity v_x $(= \partial \psi/\partial y)$ and v_y $(= -\partial \psi/\partial x)$ can be computed from Ψ_0 and are given by

$$v_x = \frac{3}{2H}\left[\left(\frac{y}{H}\right)^2 - 1\right], \qquad v_y = \frac{3\varepsilon H'}{2H}\left[\left(\frac{y}{H}\right)^3 - \left(\frac{y}{H}\right)\right] \qquad\qquad (2.2\text{-}32a,b)$$

The velocity component v_x is of zeroth order and at any given H, it has the usual parabolic profile. This solution is the quasi-cylindrical approximate solution (Van Dyke, 1987). Note that Ψ_0 also yields the correct first order (order ε) approximation for v_y.

By introducing the new scale X, we have transferred ε from the boundary conditions to the differential equations and have obtained a uniformly valid solution.

●

The extension to more than two scales is straightforward but it leads to extensive computations. The next example illustrates the case of three scales.

Example 2.2-3. **Mathieu's equation**

Solve Mathieu's equation [Equations (1.5-36, 38)] by the method of multiple scales.

Mathieu's equation can also be written as

$$\frac{d^2x}{dt^2} + (\omega^2 + \varepsilon \cos 2t)\, x = 0 \qquad (2.2\text{-}33)$$

where, for convenience, δ is replaced by ω^2.

Note that when $\varepsilon = 0$, Equation (2.2-33) is the simple harmonic equation. The term $\varepsilon x \cos 2t$ can be considered to be an imposed periodic force.

In Problem 10 of Chapter 1, we have discussed the qualitative properties of the solutions of Equation (2.2-33). In particular, we have outlined how it can be deduced that periodic solutions exist if $|\phi(\omega^2, \varepsilon)| = 2$, where ϕ is not known explicitly. These values of ω^2 and ε that yield periodic solutions separate the bounded solutions from the unbounded solutions. We now use the perturbation method to determine the values of ω^2 and ε such that $\phi = \pm 2$. To be specific, we seek solutions of period π (corresponding to $\phi = 2$) and assume that the initial conditions are

$$x(0) = 1, \qquad \dot{x}(0) = 0 \qquad (2.2\text{-}34a,b)$$

We first expand x as

$$x(t) = x_0(t) + \varepsilon x_1(t) + \dots \qquad (2.2\text{-}35)$$

Differentiating, substituting, and comparing powers of ε yields

$$\varepsilon^0: \qquad \ddot{x}_0 + \omega^2 x_0 = 0 \qquad (2.2\text{-}36)$$

The solution of Equations (2.2-34a, b, 36) is

$$x_0 = \cos \omega t \qquad (2.2\text{-}37)$$

The solution is of period π if $\omega = 2n$, where n is an integer. We seek a solution in the neighborhood of $\omega = 2$ and write ω as

$$\omega = 2 + \varepsilon \omega_1 + \varepsilon^2 \omega_2 + \dots \qquad (2.2\text{-}38)$$

where ω_1 and ω_2 are constants.

We introduce three time scales defined by

$$T_0 = t, \qquad T_1 = \varepsilon t, \qquad T_2 = \varepsilon^2 t \tag{2.2-39a,b,c}$$

The function $x(t)$ is now written as $X(T_0, T_1, T_2)$.

The chain rule yields

$$\frac{dx}{dt} = \frac{\partial X}{\partial T_0} + \varepsilon \frac{\partial X}{\partial T_1} + \varepsilon^2 \frac{\partial X}{\partial T_2} \tag{2.2-40}$$

Differentiating (dx/dt) again, Equation (2.2-33) becomes

$$\frac{\partial^2 X}{\partial T_0^2} + 2\varepsilon \frac{\partial^2 X}{\partial T_0 \partial T_1} + \varepsilon^2 \left(2 \frac{\partial^2 X}{\partial T_0 \partial T_2} + \frac{\partial^2 X}{\partial T_1^2} \right) + \dots$$

$$+ (4 + 4\varepsilon\omega_1 + 4\varepsilon^2\omega_2 + \varepsilon^2\omega_1^2 + \dots + \varepsilon\cos 2T_0) X = 0 \tag{2.2-41}$$

We expand X as

$$X = X_0 + \varepsilon X_1 + \varepsilon^2 X_2 + \dots \tag{2.2-42}$$

Differentiating, substituting the resulting expressions in Equation (2.2-41), and comparing powers of ε yields

$$\varepsilon^0: \qquad \frac{\partial^2 X_0}{\partial T_0^2} + 4X_0 = 0 \tag{2.2-43a}$$

$$\varepsilon: \qquad \frac{\partial^2 X_1}{\partial T_0^2} + 4X_1 = -\left[2 \frac{\partial^2 X_0}{\partial T_0 \partial T_1} + (4\omega_1 + \cos 2T_0) X_0 \right] \tag{2.2-43b}$$

$$\varepsilon^2: \qquad \frac{\partial^2 X_2}{\partial T_0^2} + 4X_2 = -\left[2 \frac{\partial^2 X_0}{\partial T_0 \partial T_2} + 2 \frac{\partial^2 X_1}{\partial T_0 \partial T_1} + \frac{\partial^2 X_0}{\partial T_1^2} + (4\omega_1 + \cos 2T_0) X_1 + (4\omega_2 + \omega_1^2) X_0 \right]$$

$$\tag{2.2-43c}$$

The initial conditions are

$$X_0 = 1 ; \quad X_i = 0, \ i = 1, 2 ; \quad \partial X_i / \partial T_j = 0, \ i, j = 0, 1, 2 \tag{2.2-44a,b,c}$$

where all the functions and derivatives are evaluated at $T_i = 0, \ i = 0, 1, 2$.

The solution of Equation (2.2-43a) is

$$X_0 = A_0 (T_1, T_2) \cos 2T_0 + B_0 (T_1, T_2) \sin 2T_0 \tag{2.2-45}$$

Substituting X_0 into Equation (2.2-43b) yields

$$\frac{\partial^2 X_1}{\partial T_0^2} + 4X_1 = -\left[-4\frac{\partial A_0}{\partial T_1}\sin 2T_0 + 4\frac{\partial B_0}{\partial T_1}\cos 2T_0 + 4\omega_1 A_0 \cos 2T_0 \right.$$

$$\left. + 4\omega_1 B_0 \sin 2T_0 + A_0 \cos^2 2T_0 + B_0 \cos 2T_0 \sin 2T_0 \right] \qquad (2.2\text{-}46a)$$

$$= -\left[\left(4\frac{\partial B_0}{\partial T_1} + 4\omega_1 A_0\right)\cos 2T_0 + \left(-4\frac{\partial A_0}{\partial T_1} + 4\omega_1 B_0\right)\sin 2T_0 \right.$$

$$\left. + \frac{A_0}{2}(1 + \cos 4T_0) + \frac{B_0}{2}\sin 4T_0 \right] \qquad (2.2\text{-}46b)$$

To avoid having secular terms, we require the coefficients of $\cos 2T_0$ and $\sin 2T_0$ to be zero, that is,

$$\frac{\partial A_0}{\partial T_1} - \omega_1 B_0 = 0 \qquad (2.2\text{-}47a)$$

$$\frac{\partial B_0}{\partial T_1} + \omega_1 A_0 = 0 \qquad (2.2\text{-}47b)$$

Differentiating Equation (2.2-47a) with respect to T_1 and adding the resulting expressions to Equation (2.2-47b) yields

$$\frac{\partial^2 A_0}{\partial T_1^2} + \omega_1^2 A_0 = 0 \qquad (2.2\text{-}48)$$

The solution is

$$A_0(T_1, T_2) = C_0(T_2)\cos \omega_1 T_1 + D_0(T_2)\sin \omega_1 T_1 \qquad (2.2\text{-}49)$$

From Equation (2.2-47a), we deduce that

$$B_0(T_1, T_2) = -C_0(T_2)\sin \omega_1 T_1 + D_0(T_2)\cos \omega_1 T_1 \qquad (2.2\text{-}50)$$

Combining Equations (2.2-45, 49, 50) and simplifying the trigonometric expressions yields

$$X_0 = C_0(T_2)\cos(2T_0 + \omega_1 T_1) + D_0(T_2)\sin(2T_0 + \omega_1 T_1) \qquad (2.2\text{-}51a)$$

$$= C_0 \cos(2t + \varepsilon\omega_1 t) + D_0 \sin(2t + \varepsilon\omega_1 t) \qquad (2.2\text{-}51b)$$

Imposing the initial conditions, the condition that X_0 is of period π and $\varepsilon \neq 0$ yields

$$\omega_1 = 0, \quad D_0 = 0, \quad C_0(0) = 1 \qquad (2.2\text{-}52a,b,c)$$

From Equations (2.2-45, 46b, 49, 50, 52a–c), we deduce that

$$X_0 = \cos 2T_0 \tag{2.2-53}$$

$$\frac{\partial^2 X_1}{\partial T_0^2} + 4X_1 = -\frac{1}{2}(1 + \cos 4T_0) \tag{2.2-54}$$

The solution is

$$X_1 = A_1(T_1, T_2)\cos 2T_0 + B_1(T_1, T_2)\sin 2T_0 + (1/8)[(1/3)\cos 4T_0 - 1] \tag{2.2-55}$$

Substituting X_0 and X_1 into Equations (2.2-43c) yields

$$\frac{\partial^2 X_2}{\partial T_0^2} + 4X_2 = -\left[-4\frac{\partial A_1}{\partial T_1}\sin 2T_0 + 4\frac{\partial B_1}{\partial T_1}\cos 2T_0 + \frac{A_1}{2}(1 + \cos 4T_0) + \frac{B_1}{2}\sin 4T_0 \right.$$
$$\left. + \frac{1}{8}\left(\frac{\cos 6T_0 + \cos 2T_0}{6} - \cos 2T_0\right) + 4\omega_2 \cos 2T_0 \right] \tag{2.2-56a}$$

$$= -\left[-4\frac{\partial A_1}{\partial T_1}\sin 2T_0 + \left(4\frac{\partial B_1}{\partial T_1} - \frac{5}{48} + 4\omega_2\right)\cos 2T_0 + \frac{A_1}{2}(1 + \cos 4T_0) \right.$$
$$\left. + \frac{B_1}{2}\sin 4T_0 + \frac{a \cos 6T_0}{48} \right] \tag{2.2-56b}$$

To eliminate the secular terms, we require

$$\frac{\partial A_1}{\partial T_1} = 0 \tag{2.2-57a}$$

$$4\frac{\partial B_1}{\partial T_1} - \frac{5}{48} + 4\omega_2 = 0 \tag{2.2-57b}$$

We deduce that a solution of Equations (2.2-57a, b) is

$$A_1 = B_1 = \text{constant} \tag{2.2-58a,b}$$

$$\omega_2 = 5/192 \tag{2.2-58c}$$

Substituting values of ω_1 and ω_2 into Equation (2.2-38) yields

$$\omega = 2 + (5/192)\varepsilon^2 + \ldots \tag{2.2-59a}$$

or $\quad \omega^2 = 4 + (5/48)\varepsilon^2 + \ldots \tag{2.2-59b}$

Equation (2.2-59b) expresses the parametric relationship that yields the periodic solution of period π near $\omega^2 = 4$ satisfying the initial conditions. If in Equation (2.2-53), X_0 is given by $\sin 2T_0$, Equation (2.2-59b) will not be the same.

Equation (2.2-55) now becomes

$$X_1 = A_1 \cos 2T_0 + B_1 \sin 2T_0 + (1/8) \left[(1/3) \cos 4T_0 - 1 \right] \tag{2.2-60}$$

Imposing the initial conditions $[X_1(0) = \dot{X}_1(0) = 0]$ yields

$$A_1 = 1/12, \qquad B_1 = 0 \tag{2.2-61a,b}$$

Combining Equations (2.2-42, 53, 60, 61a, b) yields

$$X = \cos 2T_0 + \varepsilon \left[(1/12) \cos 2T_0 + (1/24) \cos 4T_0 - (1/8) \right] + \dots \tag{2.2-62a}$$

or $\quad x = \cos 2t + \varepsilon \left[(1/12) \cos 2t + (1/24) \cos 4t - (1/8) \right] + \dots \tag{2.2-62b}$

●

We can extend the number of time scales T_0, T_1, T_2, ... to as many as we (or the computer) can handle and the amount of time needed to perform all the required computations can be enormous. In the **two-variable method**, we introduce only two time scales.

The slow time scale is T_1 ($= \varepsilon t$) and the fast time scale T is defined by

$$T = t \left(1 + \varepsilon^2 \omega_2 + \varepsilon^3 \omega_3 + \dots \right) \tag{2.2-63}$$

where ω_2, ω_3, ... are constants.

Note the absence of the term εt ($= T_1$) in Equation (2.2-63). The independent function $x(t)$ is now a function of two variables T and T_1. The next example illustrates the two-variable method.

Example 2.2-4. **Van der Pol's equation**

Solve the Van der Pol equation [Equation (1.4-25)] by the two-variable method.

For convenience, we reproduce Equation (1.4-25) here (with $\mu = \varepsilon$)

$$\ddot{x} - \varepsilon (1 - x^2) \dot{x} + x = 0 \tag{1.4-25}$$

In Chapter 1, we have shown that Equation (1.4-25) has a periodic solution and here we are going to compute it. We treat $x(t)$ as a function of two variables $X(T, T_1)$.

Applying the chain rule yields

$$\frac{dx}{dt} = (1 + \epsilon^2 \omega_2 + ...) \frac{\partial X}{\partial T} + \epsilon \frac{\partial X}{\partial T_1} \tag{2.2-64a}$$

$$\frac{d^2 x}{dt^2} = \frac{\partial^2 X}{\partial T^2} + 2\epsilon \frac{\partial^2 X}{\partial T \partial T_1} + \epsilon^2 \left(2\omega_2 \frac{\partial^2 X}{\partial T^2} + \frac{\partial^2 X}{\partial T_1^2} \right) \tag{2.2-64b}$$

We expand X as

$$X = X_0 (T, T_1) + \epsilon X_1 + \epsilon^2 X_2 + ... \tag{2.2-65}$$

Combining Equations (1.4-25, 2.2-64a, b, 65) and comparing powers of ϵ yields

$$\epsilon^0: \qquad \frac{\partial^2 X_0}{\partial T^2} + X_0 = 0 \tag{2.2-66a}$$

$$\epsilon: \qquad \frac{\partial^2 X_1}{\partial T^2} + X_1 = -2 \frac{\partial^2 X_0}{\partial T \partial T_1} + (1 - X_0^2) \frac{\partial X_0}{\partial T} \tag{2.2-66b}$$

$$\epsilon^2: \qquad \frac{\partial^2 X_2}{\partial T^2} + X_2 = -2 \frac{\partial^2 X_1}{\partial T \partial T_1} - 2\omega_2 \frac{\partial^2 X_0}{\partial T^2} - \frac{\partial^2 X_0}{\partial T_1^2} + (1 - X_0^2) \left(\frac{\partial X_1}{\partial T} + \frac{\partial X_0}{\partial T_1} \right) - 2 X_0 X_1 \frac{\partial X_0}{\partial T} \tag{2.2-66c}$$

The solution of Equation (2.2-66a) is

$$X_0 = A_0 (T_1) e^{iT} + \overline{A}_0 (T_1) e^{-iT} \tag{2.2-67}$$

Substituting X_0 into Equation (2.2-66b) yields

$$\frac{\partial^2 X_1}{\partial T^2} + X_1 = e^{iT} \left[-2i \frac{dA_0}{dT_1} + (1 - 2 A_0 \overline{A}_0) i A_0 + i A_0^2 \overline{A}_0 \right] - i A_0^3 e^{3iT} + \text{c.c.} \tag{2.2-68}$$

To eliminate the secular terms, we require

$$-2i \frac{dA_0}{dT_1} + (1 - 2 A_0 \overline{A}_0) i A_0 + i A_0^2 \overline{A}_0 = 0 \tag{2.2-69a}$$

or $\qquad 2 \frac{dA_0}{dT_1} - (1 - 2 A_0 \overline{A}_0) A_0 - A_0^2 \overline{A}_0 = 0 \tag{2.2-69b}$

To solve Equation (2.2-69b), we assume a solution of the form

$$A_0(T_1) = a(T_1) \exp[i\phi(T_1)] \tag{2.2-70a}$$

where a and ϕ are real quantities.

The complex conjugate of A_0 is

$$\overline{A}_0 = a(T_1) \exp[-i\phi(T_1)] \tag{2.2-70b}$$

Differentiating A_0 yields

$$\frac{dA_0}{dT_1} = \left(\frac{da}{dT_1} + ia\frac{d\phi}{dT_1}\right)\exp(i\phi) \tag{2.2-71}$$

Combining Equations (2.2-69b – 71) yields

$$2\left(\frac{da}{dT_1} + ia\frac{d\phi}{dT_1}\right) - (1 - 2a^2)a - a^3 = 0 \tag{2.2-72}$$

The real and imaginary parts of Equation (2.2-72) are

$$\frac{da}{dT_1} = \frac{a(1 - a^2)}{2} \tag{2.2-73a}$$

$$\frac{d\phi}{dT_1} = 0 \tag{2.2-73b}$$

The solutions are

$$a^2 = 1/(1 + Ke^{-T_1}) \tag{2.2-74a}$$

$$\phi = \phi_0 \tag{2.2-74b}$$

where K and ϕ_0 are constants.

Substituting a and ϕ into Equation (2.2-70a) yields

$$A_0 = \exp(i\phi_0)/(1 + Ke^{-T_1})^{1/2} \tag{2.2-75}$$

From Equations (2.2-67, 75), we deduce that, to our degree of approximation,

$$X_0 = \cos(\phi_0 + t)/[1 + K\exp(-\varepsilon t)]^{1/2} \tag{2.2-76}$$

where K and ϕ_0 are determined by the initial conditions.

Note that as $t \longrightarrow \infty$, $X_0 \longrightarrow \cos(\phi_0 + t)$, which is a periodic solution (limit cycle). The solution X_0 already includes a first correction in ε. To improve the solution, we solve Equation (2.2-66b) for X_1. The solution, in this case, is

$$X_1 = A_1(T_1) e^{iT} + (i A_0^3 / 8) e^{3iT} + \text{c.c.} \tag{2.2-77}$$

where $A_1(T_1)$ is a complex function.

To determine A_1, we substitute X_1 and X_0 into Equation (2.2-66c). The condition that X_2 should have no secular terms leads to a differential equation involving (dA_1/dT_1) which is similar to Equation (2.2-69b) but having many more terms. Details are given in Nayfeh (1977).

●

The time scale T defined by Equation (2.2-63) can be extended to a more general form by assuming that T can be expressed as an asymptotic sequence $\phi_n(\varepsilon)$ instead of powers of ε. That is to say, T is written as

$$T = t \left[1 + \sum_{n=1}^{\infty} \omega_n \phi_n(\varepsilon) \right] \tag{2.2-78}$$

where ω_n are constants. The sequence $\phi_n(\varepsilon)$ is determined as part of the problem.

Another possible extension is to express T as functions of t and in this case, we write T as

$$T = \sum_{n=0}^{\infty} \varepsilon^n g_n(t) \tag{2.2-79}$$

where the functions $g_n(t)$ are determined by the problem.

If the problem involves a boundary layer (non-uniformity near the origin), the sum in Equation (2.2-79) starts from a negative number $(n < 0)$ (see Example 2.2-1). These generalizations are discussed in Nayfeh (1973) and Kevorkian and Cole (1981).

We close this section by considering a method which involves only one time scale explicitly but it is stretched (or strained) in the following manner

$$t = s \left[1 + \sum_{n=1}^{\infty} \omega_n \varepsilon^n \right] \tag{2.2-80}$$

where ω_n are constants and s is the new independent variable.

This method is the **Lindstedt-Poincaré** method and is illustrated in the next example.

Example 2.2-5. **Lindstedt-Poincaré method**

Solve Equations (2.1-18a–c) by the Lindstedt-Poincaré method.

Form Equation (2.2-80), writing $x(t)$ as $X(s)$ and using the chain rule, we deduce that

$$\frac{d^2 x}{dt^2} = (1 + \varepsilon \omega_1 + \varepsilon^2 \omega_2 + ...)^{-2} \frac{d^2 X}{ds^2} \qquad (2.2\text{-}81)$$

Combining Equations (2.1-18a, 2-81) yields

$$\frac{d^2 X}{ds^2} + (1 + \varepsilon \omega_1 + \varepsilon^2 \omega_2 + ...)^2 (1 + \varepsilon) X = 0 \qquad (2.2\text{-}82)$$

The function $X(s)$ is expanded as

$$X(s) = X_0(s) + \varepsilon X_1(s) + \varepsilon^2 X_2(s) + ... \qquad (2.2\text{-}83)$$

Differentiating, substituting the resulting expressions in Equation (2.2-82), and comparing powers of ε yields

$$\varepsilon^0: \quad \frac{d^2 X_0}{ds^2} + X_0 = 0 \qquad (2.2\text{-}84a)$$

$$\varepsilon: \quad \frac{d^2 X_1}{ds^2} + X_1 = -(1 + 2\omega_1) X_0 \qquad (2.2\text{-}84b)$$

$$\varepsilon^2: \quad \frac{d^2 X_2}{ds^2} + X_2 = -[(1 + 2\omega_1) X_1 + (\omega_1^2 + 2\omega_1 + 2\omega_2) X_0] \qquad (2.2\text{-}84c)$$

The initial conditions [Equations (2.1-18b, c)] become

$$X_0(0) = 1, \qquad X_1(0) = X_2(0) = ... = 0 \qquad (2.2\text{-}85a,b,c)$$

$$(dX_i/ds) = 0, \quad i = 0, 1, 2, ... \qquad (2.2\text{-}85d)$$

The solution of Equation (2.2-84a), subject to the initial conditions, is

$$X_0 = \cos s \qquad (2.2\text{-}86)$$

Substituting X_0 into Equation (2.2-84b) yields

$$\frac{d^2 X_1}{ds^2} + X_1 = (1 + 2\omega_1)\, a \cos s \qquad (2.2\text{-}87)$$

To eliminate the secular term, we require that

$$\omega_1 = -1/2 \qquad (2.2\text{-}88)$$

To the order of ε, Equation (2.2-80) becomes

$$t = s\,(1 - \varepsilon/2) \qquad (2.2\text{-}89\text{a})$$

or $\qquad s = (1 - \varepsilon/2)^{-1}\, t \approx (1 + \varepsilon/2)\, t \qquad (2.2\text{-}89\text{b,c})$

Substituting Equation (2.2-89c) into Equation (2.2-86) yields

$$x_0 = \cos (1 + \varepsilon/2)\, t \qquad (2.2\text{-}90)$$

Equation (2.2-90) is identical to Equation (2.2-12b).

With $\omega_1 = -1/2$, the solution of Equation (2.2-84b) that satisfies the initial conditions is

$$X_1 = 0 \qquad (2.2\text{-}91)$$

From Equation (2.2-84c), we deduce that the condition for X_2 to be free of secular terms is

$$\omega_1^2 + 2\omega_1 + 2\omega_2 = 0 \qquad (2.2\text{-}92\text{a})$$

or $\qquad \omega_2 = -(\omega_1^2 + 2\omega_1)/2 = 3/8 \qquad (2.2\text{-}92\text{b,c})$

Substituting Equation (2.2-92c) into Equation (2.2-80) yields

$$s = (1 - \varepsilon/2 + 3\varepsilon^2/8)^{-1}\, t \qquad (2.2\text{-}93\text{a})$$

$$\approx (1 + \varepsilon/2 - \varepsilon^2/8)\, t \qquad (2.2\text{-}93\text{b})$$

Equation (2.2-93b) is identical to Equation (2.1-25a) when expanded to $O(\varepsilon^2)$. The solution can be written as

$$x \approx \cos (1 + \varepsilon/2 - \varepsilon^2/8)\, t \qquad (2.2\text{-}94)$$

Several modifications of Equation (2.2-80) have been proposed and some of them are discussed in the books on perturbations referred to earlier. These methods are the strained coordinates. Chan Man Fong et al. (1996) have shown, by solving the flow of viscoelastic fluid over a porous plate, that the problem of the availability of boundary conditions in non-Newtonian fluid mechanics can be solved by introducing strained coordinates.

In the next section, we introduce the method of averaging which is an alternative to the method of two time scales.

2.3 METHOD OF AVERAGING

This method is used to solve equations of the form

$$\frac{d^2 x}{dt^2} + \omega^2 x = \varepsilon f\left(x, \frac{dx}{dt}\right) \qquad (2.3\text{-}1)$$

If $\varepsilon = 0$, the solution is

$$x = a \cos(\omega t + \phi) \qquad (2.3\text{-}2)$$

where a and ϕ are constants.

When $\varepsilon \neq 0$, we look for an approximate solution of the form

$$x(t) = a(t) \cos[\omega t + \phi(t)] \qquad (2.3\text{-}3)$$

The present method follows closely the method of variation of parameters. The constant a and ϕ in Equation (2.3-2) are replaced by $a(t)$ and $\phi(t)$. Differentiating yields

$$\frac{dx}{dt} = -a(t)[\omega + \dot{\phi}]\sin[\omega t + \phi(t)] + \dot{a}(t)\cos[\omega t + \phi(t)] \qquad (2.3\text{-}4)$$

We now impose the condition

$$-a\dot{\phi}\sin(\omega t + \phi) + \dot{a}\cos(\omega t + \phi) = 0 \qquad (2.3\text{-}5)$$

From Equations (2.3-4, 5), we deduce that

$$\frac{d^2 x}{dt^2} = -a\omega^2\cos(\omega t + \phi) - a\omega\dot{\phi}\cos(\omega t + \phi) - \dot{a}\omega\sin(\omega t + \phi) \qquad (2.3\text{-}6)$$

Substituting x and $(d^2 x / dt^2)$ into Equation (2.3-1) yields

$$a\omega\dot{\phi}\cos(\omega t+\phi)+\dot{a}\omega\sin(\omega t+\phi)=-\varepsilon f[a\cos(\omega t+\phi),-a\omega\sin(\omega t+\phi)] \qquad (2.3\text{-}7)$$

Equation (2.3-7) shows that $\dot{a}$ and $\dot{\phi}$ are of $O(\varepsilon)$ and are slowly varying functions of time. Various techniques have been proposed to replace $\dot{a}$ and $\dot{\phi}$ by their average quantities. In the next example, we introduce the **Krylov-Bogoliubov (1947) method**.

Example 2.3-1. **Krylov-Bogoliubov method**

Solve the Van der Pol equation by the Krylov-Bogoliubov method.

The Van der Pol equation is given by Equation (1.4-25) and we identify $f(x,\dot{x})$ as

$$f(x,\dot{x})=(1-x^2)\dot{x} \qquad (2.3\text{-}8)$$

Combining Equations (2.3-3, 4, 5, 8), Equation (2.3-7) becomes

$$a\omega\dot{\phi}\cos\alpha+\dot{a}\omega\sin\alpha=\varepsilon[a\omega\sin\alpha-a^3\omega\cos^2\alpha\sin\alpha] \qquad (2.3\text{-}9)$$

where $\alpha=\omega t+\phi$.

Solving for $\dot{a}$ and $\dot{\phi}$ from Equations (2.3-5, 9) yields

$$\dot{a}=a\varepsilon\sin^2\alpha\,(1-a^2\cos^2\alpha) \qquad (2.3\text{-}10\text{a})$$

$$\dot{\phi}=\varepsilon\sin\alpha\cos\alpha\,(1-a^2\cos^2\alpha) \qquad (2.3\text{-}10\text{b})$$

From Equations (2.3-2), we deduce that x is of period $2\pi/\omega$ and since $\dot{a}$ and $\dot{\phi}$ are slowly varying functions, we assume that a and ϕ are constant over one period. The right side of Equations (2.3-10a, b) are replaced by their averages over a period.

Equations (2.3-10a, b) now become

$$\dot{a}=\frac{a\varepsilon}{(2\pi/\omega)}\int_0^{2\pi/\omega}[\sin^2(\omega t+\phi)]\,[1-a^2\cos^2(\omega t+\phi)]\,dt \qquad (2.3\text{-}11\text{a})$$

$$=\frac{a\varepsilon}{2\pi}\int_0^{2\pi}\sin^2\alpha\,(1-a^2\cos^2\alpha)\,d\alpha \qquad (2.3\text{-}11\text{b})$$

$$= \frac{a\varepsilon}{2} \left(1 - \frac{a^2}{4} \right)$$

(2.3-11c)

$$\dot{\phi} = \frac{\varepsilon\omega}{2\pi} \int_0^{2\pi/\omega} [\sin(\omega t + \phi) \cos(\omega t + \phi)] [1 - a^2\cos^2(\omega t + \phi)] \, dt$$

(2.3-12a)

$$= \frac{\varepsilon\omega}{2\pi} \int_0^{2\pi} [\sin\alpha \cos\alpha] [1 - a^2\cos^2\alpha] \, d\alpha$$

(2.3-12b)

$$= 0$$

(2.3-12c)

The solutions of Equations (2.3-11c, 12c) are

$$a^2 = 4 / [1 + (4/K) e^{-\varepsilon t}]$$

(2.3-13a)

$$\phi = \phi_0$$

(2.3-13b)

where K and ϕ_0 are constants.

From Equations (2.3-3, 13a, b), we deduce that

$$x(t) = \frac{2}{[1 + (4/K) e^{-\varepsilon t}]^{1/2}} \cos(\omega t + \phi_0)$$

(2.3-14)

Equation (2.3-14) is of the same form as Equation (2.2-76) with $\omega = 1$.

●

Example 2.3-2. **Relativistic orbital equation**

The orbital equation of a planet about the sun, when the effect of relativity is included, can be written as

$$\frac{d^2 u}{d\theta^2} + u = k(1 + \varepsilon u^2)$$

(2.3-15)

where $u = (1/r)$, r and θ are polar coordinates, $k = \gamma m/h^2$, γ is the gravitational constant, m is the mass of the planet, h is the moment of momentum which is a constant, and $\varepsilon k u^2$ is the relativistic correction term.

Solve Equation (2.3-15) by the method of Krylov and Bogoliubov, subject to the initial conditions

$$u(0) = k(1 + e), \qquad \dot{u}(0) = 0$$

(2.3-16a,b)

where e is the eccentricity of the unperturbed orbit.

In this example, the independent variable is θ and not t and when $\varepsilon = 0$, Equation (2.3-15) is non-homogeneous. In this case, we assume

$$u(\theta) = a(\theta) \cos[\theta + \phi(\theta)] + k \qquad (2.3\text{-}17)$$

Equation (2.3-5) now reads

$$-a \frac{d\phi}{d\theta} \sin(\theta + \phi) + \frac{da}{d\theta} \cos(\theta + \phi) = 0 \qquad (2.3\text{-}18)$$

It follows from Equations (2.3-17, 18) that

$$\frac{d^2 u}{d\theta^2} = -a \cos(\theta + \phi) - a \frac{d\phi}{d\theta} \cos(\theta + \phi) - \frac{da}{d\theta} \sin(\theta + \phi) \qquad (2.3\text{-}19)$$

Combining Equations (2.3-15, 17, 19) yields

$$a \frac{d\phi}{d\theta} \cos(\theta + \phi) + \frac{da}{d\theta} \sin(\theta + \phi) = -k\varepsilon [a^2 \cos^2(\theta + \phi) + 2ak \cos(\theta + \phi) + k^2] \qquad (2.3\text{-}20)$$

Solving for $(da/d\theta)$ and $(d\phi/d\theta)$ from Equations (2.3-18, 20) yields

$$\frac{da}{d\theta} = -k\varepsilon \sin\alpha \, [a^2 \cos^2\alpha + 2ak \cos\alpha + k^2] \qquad (2.3\text{-}21\text{a})$$

$$\frac{d\phi}{d\theta} = -\frac{k\varepsilon \cos\alpha}{a} \, [a^2 \cos^2\alpha + 2ak \cos\alpha + k^2] \qquad (2.3\text{-}21\text{b})$$

where $\alpha = \theta + \phi$.

From Equations (2.3-21a, b), we deduce that $(da/d\theta)$ and $(d\phi/d\theta)$ are slowly varying functions and we can approximate the terms on the right side of Equations (2.3-21a, b) by their average values over a period of 2π. Over a period of 2π, a and ϕ are assumed to be constants. Equations (2.3-21a, b) simplify to

$$\frac{da}{d\theta} = 0 \qquad (2.3\text{-}22\text{a})$$

$$\frac{d\phi}{d\theta} = -\varepsilon k^2 \qquad (2.3\text{-}22\text{b})$$

The solution is

$$a = a_0, \qquad \phi = -\varepsilon k^2 \theta + \phi_0 \qquad\qquad (2.3\text{-}23a,b)$$

where a_0 and ϕ_0 are constants.

Substituting a and ϕ into Equations (2.3-17) yields

$$u(\theta) = a_0 \cos[(1 - \varepsilon k^2)\theta + \phi_0] + k \qquad\qquad (2.3\text{-}24)$$

From the initial conditions [Equations (2.3-16a, b)], we deduce that

$$a_0 = ke, \qquad \phi_0 = 0 \qquad\qquad (2.3\text{-}25a,b)$$

The function u is

$$u(\theta) = k[e\cos(1 - \varepsilon k^2)\theta + 1] \qquad\qquad (2.3\text{-}26)$$

At the perihelion (the nearest point to the sun on the unperturbed orbit), u is a maximum (r is a minimum) and is given by

$$\cos(1 - \varepsilon k^2)\theta = 1 \qquad\qquad (2.3\text{-}27a)$$

or $\qquad (1 - \varepsilon k^2)\theta = 2n\pi \qquad\qquad (2.3\text{-}27b)$

where n is an integer.

It follows from Equation (2.3-27b) that θ is given by

$$\theta = 2n\pi(1 - \varepsilon k^2)^{-1} \qquad\qquad (2.3\text{-}28a)$$

or $\qquad \approx 2n\pi(1 + \varepsilon k^2) \qquad\qquad (2.3\text{-}28b)$

We deduce from Equation (2.3-28b) that after each orbit, the perihelion advances by $2\pi\varepsilon k^2$. By substituting the appropriate values of ε and k for the planet Mercury, it is found that its perihelion advances by 43" per century due to the effect of relativity. This is in agreement with observation and confirms Einstein's relativity theory. For further details, see Pollard (1966).

●

The Krylov-Bogoliubov method provides the solution to the first order of ε. To obtain the higher order terms in ε, we use the Krylov-Bogoliubov-Mitropolski (KBM) method. In this method, the solution of Equation (2.3-1) is written as

$$x = a(t) \cos \alpha(t) + \sum_{n=1}^{N} \varepsilon^n x_n(a, \alpha) \qquad (2.3\text{-}29)$$

where $\alpha(t) = \omega t + \phi$ and x_n is a periodic function in α of period 2π.

We have shown earlier that a and ϕ are slowly varying functions and we can assume that

$$\frac{da}{dt} = \varepsilon A_1(a) + \varepsilon^2 A_2(a) + \dots \qquad (2.3\text{-}30a)$$

$$\frac{d\alpha}{dt} = \omega + \frac{d\phi}{dt} = \omega + \varepsilon B_1(a) + \varepsilon^2 B_2(a) + \dots \qquad (2.3\text{-}30b)$$

Differentiating x in Equation (2.3-29) with respect to t yields

$$\frac{dx}{dt} = \frac{da}{dt} \cos \alpha(t) - a \frac{d\alpha}{dt} \sin \alpha(t) + \sum_{n=1}^{N} \varepsilon^n \left[\frac{\partial x_n}{\partial a} \frac{da}{dt} + \frac{\partial x_n}{\partial \alpha} \frac{d\alpha}{dt} \right] \qquad (2.3\text{-}31a)$$

$$\frac{d^2 x}{dt^2} = \frac{d^2 a}{dt^2} \cos \alpha - 2 \frac{da}{dt} \frac{d\alpha}{dt} \sin \alpha - a \frac{d^2 \alpha}{dt^2} \sin \alpha - a \left(\frac{d\alpha}{dt} \right)^2 \cos \alpha$$

$$+ \sum_{n=1}^{N} \varepsilon^n \left[\frac{\partial x_n}{\partial a} \frac{d^2 a}{dt^2} + \frac{\partial x_n}{\partial \alpha} \frac{d^2 \alpha}{dt^2} + \frac{\partial^2 x_n}{\partial a^2} \left(\frac{da}{dt} \right)^2 + 2 \frac{\partial^2 x_n}{\partial \alpha \partial a} \left(\frac{da}{dt} \right) \left(\frac{d\alpha}{dt} \right) + \frac{\partial^2 x_n}{\partial \alpha^2} \left(\frac{d\alpha}{dt} \right)^2 \right]$$

$$(2.3\text{-}31b)$$

Combining Equations (2.3-1, 29 – 31b) and comparing powers of ε yields a set of equations that allows us to determine x. The next example uses this method to solve the Van der Pol equation.

Example 2.3-3. **Method of KBM**

Solve the Van der Pol equation [Equation (1.4-25)] by the KBM method.

The quantities $(d^2 a/dt^2)$ and $(d^2 \alpha/dt^2)$ can be derived from Equations (2.3-30a, b) and they are

$$\frac{d^2 a}{dt^2} = \varepsilon \frac{dA_1}{da} \frac{da}{dt} + \varepsilon^2 \frac{dA_2}{da} \frac{da}{dt} + \dots \qquad (2.3\text{-}32a)$$

$$= \varepsilon^2 A_1 \frac{dA_1}{da} + \varepsilon^3 \left(A_2 \frac{dA_1}{da} + A_1 \frac{dA_2}{da} \right) + \dots \qquad (2.3\text{-}32b)$$

$$\frac{d^2 \alpha}{dt^2} = \varepsilon \frac{dB_1}{da} \frac{da}{dt} + \varepsilon^2 \frac{dB_2}{da} \frac{da}{dt} + \dots \qquad (2.3\text{-}32c)$$

$$= \varepsilon^2 A_1 \frac{dB_1}{da} + \varepsilon^3 \left(A_2 \frac{dB_1}{da} + A_1 \frac{dB_2}{da}\right) + \dots \qquad (2.3\text{-}32d)$$

Substituting Equations (2.3-30a, b, 32b, d) into Equations (2.3-31a, b) yields

$$\frac{dx}{dt} = -a \sin \alpha + \varepsilon \left(A_1 \cos \alpha + \frac{\partial x_1}{\partial \alpha} - a B_1 \sin \alpha\right) + \varepsilon^2 \left(A_2 \cos \alpha + A_1 \frac{\partial x_1}{\partial a} + \frac{\partial x_2}{\partial \alpha}\right.$$

$$\left. + B_1 \frac{\partial x_1}{\partial \alpha} - a B_2 \sin \alpha\right) + \dots \qquad (2.3\text{-}33a)$$

$$\frac{d^2 x}{dt^2} = -a \cos \alpha + \varepsilon \left(\frac{\partial^2 x_1}{\partial \alpha^2} - 2 a B_1 \cos \alpha - 2 A_1 \sin \alpha\right) + \varepsilon^2 \left[2 A_1 \frac{\partial^2 x_1}{\partial a \partial \alpha} + 2 B_1 \frac{\partial^2 x_1}{\partial \alpha^2} + \frac{\partial^2 x_2}{\partial \alpha^2}\right.$$

$$\left. + \cos \alpha \left(A_1 \frac{dA_1}{da} - 2 a B_2 - a B_1^2\right) - \sin \alpha \left(a A_1 \frac{dB_1}{da} + 2 A_1 B_1 + 2 A_2\right)\right] \qquad (2.3\text{-}33b)$$

Combining Equations (1.4-25, 2.3-29, 33a, b) and comparing powers of ε yields

$$\varepsilon: \quad \frac{\partial^2 x_1}{\partial \alpha^2} + x_1 = -a \sin \alpha + a^3 \sin \alpha \cos^2 \alpha + 2 a B_1 \cos \alpha + 2 A_1 \sin \alpha \qquad (2.3\text{-}34a)$$

$$= 2 A_1 \sin \alpha + 2 a B_1 \cos \alpha - a(1 - a^2/4) \sin \alpha + (a^3/4) \sin 3\alpha \qquad (2.3\text{-}34b)$$

$$\varepsilon^2: \quad \frac{\partial^2 x_2}{\partial \alpha^2} + x_2 = \left[a(2 B_2 + B_1^2) - A_1 \frac{dA_1}{da}\right] \cos \alpha + \left[2(A_2 + A_1 B_1) + a A_1 \frac{dB_1}{da}\right] \sin \alpha - 2 B_1 \frac{\partial^2 x_1}{\partial \alpha^2}$$

$$- 2 A_1 \frac{\partial^2 x_1}{\partial a \partial \alpha} + (1 - a^2 \cos^2 \alpha) \left(A_1 \cos \alpha - a B_1 \sin \alpha + \frac{\partial x_1}{\partial \alpha}\right) + a^2 x_1 \sin 2\alpha \quad (2.3\text{-}34c)$$

By the choice of x [Equation (2.3-29)], the zeroth order equation is automatically satisfied.

To eliminate the secular terms, we require the coefficients of $\sin \alpha$ and $\cos \alpha$ in Equation (2.3-34b) to be zero, that is,

$$B_1 = 0, \qquad A_1 = \frac{a}{2}\left(1 - \frac{a^2}{4}\right) \qquad (2.3\text{-}35a,b)$$

Substituting values of A_1 and B_1 into Equation (2.3-30a, b) yields

$$\frac{da}{dt} = \varepsilon \frac{a}{2}\left(1 - \frac{a^2}{4}\right) + \dots \qquad (2.3\text{-}36a)$$

$$\frac{d\alpha}{dt} = 1 + O(\varepsilon^2) \qquad (2.3\text{-}36b)$$

Equations (2.3-36a, b) are identical to Equations (2.3-11c, 12c) and both methods generate the same solution. However, in the KBM method, we can compute the higher order approximations and we proceed to obtain the corrections of $O(\varepsilon^2)$. With the values of A_1 and B_1 given by Equations (2.3-35a, b), the solution of Equation (2.3-34b) is

$$x_1 = -(a^3/32) \sin 3\alpha \tag{2.3-37}$$

Substituting the values of B_1 $(= 0)$ and x_1 into Equation (2.3-34c) yields

$$\frac{\partial^2 x_2}{\partial \alpha^2} + x_2 = \left[2aB_2 - A_1 \frac{dA_1}{da} + (1 - \frac{3}{4} a^2) A_1 + \frac{a^5}{128} \right] \cos \alpha + 2A_2 \sin \alpha$$

$$+ \frac{a^3 (a^2 + 8)}{128} \cos 3\alpha + \frac{5a^5}{128} \cos 5\alpha \tag{2.3-38}$$

The coefficients of $\sin \alpha$ and $\cos \alpha$ must be zero, so that x_2 has no secular terms, that is,

$$A_2 = 0 , \qquad B_2 = \frac{1}{2a} \left[A_1 \frac{dA_1}{da} + (1 - \frac{3}{4} a^2) A_1 + \frac{a^5}{128} \right] \tag{2.3-39a,b}$$

To $O(\varepsilon^2)$, a is still given by Equation (2.3-36b) and α is obtained by solving

$$\frac{d\alpha}{dt} = 1 + \varepsilon^2 B_2 \tag{2.3-40}$$

B_2 can be expressed as a function of t by combining Equations (2.3-13a, 35b) and α can be determined. The details can be found in Nayfeh (1973).

●

The KBM method appears to be complicated with several changes of variables but once (dx/dt) and (d^2x/dt^2) have been computed, as in Equations (2.3-33a, b), the procedure is straightforward. We demonstrate this in the next example.

Example 2.3-4. **Application of KBM method**

Solve Equation (2.1-18a) by the KBM method.

We rewrite Equation (2.1-18a) in the form of Equation (2.3-1) as

$$\frac{d^2 x}{dt^2} + x = -\varepsilon x \tag{2.1-18a}$$

Substituting Equations (2.3-29, 33b) into Equation (2.1-18a) and comparing powers of ε yields

ε: $$\frac{d^2 x_1}{d\alpha^2} + x_1 = a(2B_1 - 1)\cos\alpha + 2A_1 \sin\alpha \tag{2.3-41a}$$

ε^2: $$\frac{d^2 x_2}{d\alpha^2} + x_2 = -x_1 - 2A_1 \frac{\partial^2 x_1}{\partial a\, \partial\alpha} - 2B_1 \frac{\partial^2 x_1}{\partial\alpha^2} + \left[aB_1^2 + 2aB_2 - A_1 \frac{dA_1}{da}\right]\cos\alpha$$

$$+ \left[a A_1 \frac{dB_1}{da} + 2A_1 B_1 + 2A_2\right]\sin\alpha \tag{2.3-41b}$$

The absence of secular terms in x_1 implies that

$$A_1 = 0, \qquad B_1 = 1/2 \tag{2.3-42a,b}$$

Substituting Equations (2.3-42a, b) into Equations (2.3-30a, b) yields

$$\frac{da}{dt} = 0 \quad \Rightarrow \quad a = \text{constant } (c_0) \tag{2.3-43a,b}$$

$$\frac{d\alpha}{dt} = 1 + \varepsilon/2 \quad \Rightarrow \quad \alpha = (1 + \varepsilon/2)\, t + \phi_0 \tag{2.3-43c,d}$$

The solution to $O(\varepsilon)$ is

$$x = c_0 \cos\left[(1 + \varepsilon/2)\, t + \phi_0\right] \tag{2.3-44}$$

Imposing the initial conditions [Equations (2.1-18b, c)] yields Equation (2.2-90) [that is, $c_0 = a = 1$, $\phi_0 = 0$]. With the values of A_1 and B_1 given by Equations (2.3.42a, b), the solution of Equation (2.3-41a), subject to the initial conditions, is

$$x_1 = 0 \tag{2.3-45}$$

Substituting the values of A_1, B_1, a, and x_1 into Equation (2.3-41b) yields

$$\frac{d^2 x_2}{d\alpha^2} + x_2 = \left(\frac{1}{4} + 2B_2\right)\cos\alpha + 2A_2 \sin\alpha \tag{2.3-46}$$

The absence of secular terms in x_2 implies

$$B_2 = -1/8, \qquad A_2 = 0 \tag{2.3-47a,b}$$

To order ε^2, a is a constant and from Equation (2.3-30b), we deduce that α is given by

$$\alpha = (1 + \varepsilon/2 - \varepsilon^2/8)\, t + \phi_0 \tag{2.3-48}$$

Imposing the initial conditions leads to

$$x = \cos\left[(1 + \varepsilon/2 - \varepsilon^2/8)\,t\right] + O\,(\varepsilon^3) \tag{2.3-49}$$

Equation (2.3-49) is identical to Equation (2.2-94).

●

Minorsky (1962) has solved several problems associated with autonomous and non-autonomous systems by the method of averaging. He has also described briefly the developments of the various versions of this method.

The next section introduces the method of matched asymptotic expansions.

2.4   MATCHED ASYMPTOTIC EXPANSIONS

In Section 2, we have stated that the straightforward expansion generates a solution that is valid only in a certain region. To obtain a uniformly valid solution, we have introduced another time scale T_1. In the method of multiple scales, the dependent function $x\,(t)$ is then considered to be a function of at least two variables T_0 ($= t$) and T_1. In the method of matched asymptotic expansions, we assume that in one region, the appropriate time scale is T_1 and the dependent variable is a function of T_1 only. This region is the **inner region** or **boundary layer**. In the other region, where the regular perturbation generates a valid solution, the independent variable is t and the dependent variable is a function of t. This region is the **outer region**. There is an overlapping region where both solutions are valid. Any undetermined constants are evaluated by matching the two solutions in this overlapping region. This method was first introduced to solve flow problems. Van Dyke (1975) has described the development of this method and has given many examples. We illustrate this method by considering a few examples from various fields of applied mathematics.

Example 2.4-1. **Michaelis-Menten equation – revisited**

Solve Equations (2.1-32a, b, 33a, b) by the method of matched asymptotic expansions.

In Examples 2.1-3 and 2-1, we have deduced that the solution generated by the regular perturbation is not valid near $\tau = 0$. To obtain a uniformly valid solution, we have introduced a second time scale T_1 ($= \tau/\varepsilon$) [Equation (2.2-13b)]. Near $\tau = 0$, the appropriate time scale is T_1 and the dependent variables $u\,(\tau)$ and $v\,(\tau)$ are now denoted by $U\,(T_1)$ and $V\,(T_1)$.

Equations (2.1-32a, b) now become

$$\frac{dU}{dT_1} = -\varepsilon U + \varepsilon\,(U + \lambda_1)\,V \tag{2.4-1a}$$

$$\frac{dV}{dT_1} = U - (U + \lambda_2)\,V \tag{2.4-1b}$$

These equations are valid at $\tau = 0$ and we can apply the initial conditions [Equations (2.1-33a, b)]

$$U(0) = 1, \qquad V(0) = 0 \tag{2.4-2a,b}$$

We expand U and V as

$$U = U_0^{(i)} + \varepsilon U_1^{(i)} + \dots \tag{2.4-3a}$$

$$V = V_0^{(i)} + \varepsilon V_1^{(i)} + \dots \tag{2.4-3b}$$

The superscript (i) is to denote "inner solution".

Differentiating U and V, substituting the resulting expressions in Equations (2.4-1a, b) and comparing powers of ε yields

$$\frac{dU_0^{(i)}}{dT_1} = 0 \tag{2.4-4a}$$

$$\frac{dV_0^{(i)}}{dT_1} = U_0^{(i)} - (U_0^{(i)} + \lambda_2) V_0^{(i)} \tag{2.4-4b}$$

Equations (2.4-4a, b) are identical to Equations (2.2-16a, b) and the solutions, subject to the initial conditions, are

$$U_0^{(i)} = 1, \qquad V_0^{(i)} = [1 - \exp - (1 + \lambda_2) T_1] / (1 + \lambda_2) \tag{2.4-5a,b}$$

In the outer region, the independent variable is τ and Equations (2.1-32a, b) remain unchanged. However, the initial conditions can no longer be imposed because in the outer region $\tau \gg 0$ and we cannot impose a condition that is valid at $\tau = 0$. In this example, there is no condition to be associated with Equations (2.1-32a, b). We expand u and v as in Equations (2.1-34a, b) and denote u_0, v_0 by $u_0^{(0)}$, $v_0^{(0)}$ to indicate outer solutions. The solutions are given by Equations (2.1-37, 39) and, for convenience, they are reproduced here.

$$v_0^{(0)} = u_0^{(0)} / (u_0^{(0)} + \lambda_2) \tag{2.1-37}$$

$$(u_0^{(0)} + \lambda_2 \ell n \, u_0^{(0)}) = (\lambda_1 - \lambda_2) \tau + C_0 \tag{2.1-39}$$

The constant C_0 is to be determined by the process of matching. **Prandtl's matching principle** is the simplest and can be stated as

the inner limit of the outer solution = the outer limit of the inner solution.

In this example, applying the matching principle implies

$$\lim_{\tau \to 0} (u_0^{(0)}, v_0^{(0)}) = \lim_{T_1 \to \infty} (U_0^{(i)}, V_0^{(i)}) \tag{2.4-6}$$

From Equations (2.4-5a, b), we deduce that

$$\lim_{T_1 \to \infty} (U_0^{(i)}, V_0^{(i)}) = [1, 1/(1 + \lambda_2)] \tag{2.4-7}$$

Combining Equations (2.4-6, 7) yields

$$u_0^{(0)}(0) = 1, \qquad v_0^{(0)}(0) = 1/(1 + \lambda_2) \tag{2.4-8a,b}$$

We deduce from Equations (2.1-39, 4-8a) that

$$C_0 = 1 \tag{2.4-8c}$$

and Equation (2.4-8b) is automatically satisfied.

Substituting the value of C_0 in Equation (2.1-39) yields Equation (2.1-40). We have derived $(u_0^{(0)}, v_0^{(0)})$ which are valid for $\tau \gg 0$ and $(U_0^{(i)}, V_0^{(i)})$ which are valid for $\tau \approx 0$. We now generate the **composite solution** $(u_0^{(c)}, v_0^{(c)})$ which is valid for all values of τ.

The functions $(u_0^{(0)}, U_0^{(i)})$ can be written as

$$u_0^{(0)} = u_0^{(00)} + u_0^{(0L)} \tag{2.4-9a}$$

$$U_0^{(i)} = U_0^{(ii)} + U_0^{(0L)} \tag{2.4-9b}$$

where $u_0^{(00)} \to 0$ as $\tau \to 0$, $U_0^{(ii)} \to 0$ as $T_1 \to \infty$ (or $\varepsilon \to 0$), $u_0^{(0L)}$ and $U_0^{(0L)}$ are the values of $u_0^{(0)}$ and $U_0^{(i)}$ respectively in the overlap region. The matching principle [Equation (2.4-6)] implies

$$u_0^{(0L)} = U_0^{(0L)} \tag{2.4-10}$$

The composite solution $u_0^{(c)}$ is

$$u_0^{(c)} = u_0^{(0)} + U_0^{(i)} - U_0^{(0L)} = u_0^{(0)} \tag{2.4-11a,b}$$

Equation (2.4-11b) is obtained from Equations (2.4-5a, 7).

Note that the composite solution is the sum of the inner and outer solution minus the value of $U_0^{(i)}$ (or $u_0^{(0)}$) in the overlap region. We need to subtract once $U_0^{(0L)}$ to avoid counting it twice (once in $u_0^{(0)}$ and once in $U_0^{(i)}$).

Similarly, $v_0^{(c)}$ is given by

$$v_0^{(c)} = V_0^{(i)} + v_0^{(0)} - V_0^{(0L)} \tag{2.4-12a}$$

$$= [1 - \exp - (1 + \lambda_2) \tau/\varepsilon]/(1 + \lambda_2) + u_0^{(0)}/(\lambda_2 + u_0^{(0)}) - 1/(1 + \lambda_2) \tag{2.4-12b}$$

$$= u_0^{(0)}/(\lambda_2 + u_0^{(0)}) - \exp - (1 + \lambda_2) \tau/\varepsilon \tag{2.4-12c}$$

Note that $v_0^{(c)}$ satisfies the initial condition whereas $v_0^{(0)}$ does not. The functions $(u_0^{(c)}, v_0^{(c)})$ are the uniformly valid solutions of Equations (2.1-32a, b).

●

It was mentioned earlier that, in fluid dynamics, there is a dimensionless number, the Reynolds number Re, and we can seek solutions for high and low Reynolds number. In the next two examples, we deduce the appropriate equations for Re $\longrightarrow \infty$ and Re $\longrightarrow 0$ respectively.

Example 2.4-2. **Boundary-layer equations**

Deduce the boundary-layer equations for the flow of a Newtonian fluid past a flat plane.

Let the flat plane be given by $y = 0$ and the flow is in the x-direction. The Navier-Stokes equation for a steady two-dimensional flow in dimensionless form can be written as

$$v_x \frac{\partial v_x}{\partial x} + v_y \frac{\partial v_x}{\partial y} = -\frac{\partial p}{\partial x} + \frac{1}{Re} \left(\frac{\partial^2 v_x}{\partial x^2} + \frac{\partial^2 v_x}{\partial y^2} \right) \tag{2.4-13a}$$

$$v_x \frac{\partial v_y}{\partial x} + v_y \frac{\partial v_y}{\partial y} = -\frac{\partial p}{\partial y} + \frac{1}{Re} \left(\frac{\partial^2 v_y}{\partial x^2} + \frac{\partial^2 v_y}{\partial y^2} \right) \tag{2.4-13b}$$

where v_x and v_y are the x and y components of the velocity, p is the pressure, Re $= (\rho UL/\mu)$ is the Reynolds number, ρ is the density, μ is the viscosity, U is a typical velocity, and L is a typical length.

The equation of continuity is

$$\frac{\partial v_x}{\partial x} + \frac{\partial v_y}{\partial y} = 0 \tag{2.4-14}$$

The boundary conditions can be assumed to be

$$v_x(x, 0) = v_y(x, 0) = 0 \tag{2.4-15a,b}$$

$$\lim_{y \to \infty} v_x(x, y) = v_\infty, \qquad \lim_{y \to \infty} v_y(x, y) = 0 \qquad \qquad (2.4\text{-}15c,d)$$

where v_∞ is the free stream velocity.

We consider the case when $Re \gg 1$ and write $\varepsilon = 1/Re$.

Expanding v_x, v_y, and p as

$$v_x = v_{x0} + \varepsilon v_{x1} + \dots \qquad \qquad (2.4\text{-}16a)$$

$$v_y = v_{y0} + \varepsilon v_{y1} + \dots \qquad \qquad (2.4\text{-}16b)$$

$$p = p_0 + \varepsilon p_1 + \dots \qquad \qquad (2.4\text{-}16c)$$

and proceeding in the usual manner yields

$$v_{x0} \frac{\partial v_{x0}}{\partial x} + v_{y0} \frac{\partial v_{y0}}{\partial y} = -\frac{\partial p_0}{\partial x} \qquad \qquad (2.4\text{-}17a)$$

$$v_{x0} \frac{\partial v_{y0}}{\partial x} + v_{y0} \frac{\partial v_{y0}}{\partial y} = -\frac{\partial p_0}{\partial y} \qquad \qquad (2.4\text{-}17b)$$

Note that Equations (2.4-17a, b) are one order lower than Equations (2.4-13a, b) and (v_{x0}, v_{y0}) cannot satisfy all the boundary conditions. The source of singularity is similar to the one discussed in Example 2.1-3, namely the coefficient of the highest derivative is the small parameter ε. Equations (2.4-17a, b) are the equations of motion of an inviscid fluid and it can be deduced from these equations that the drag on a solid boundary is zero. This result (d'Alembert's paradox) is not in accord with physical reality. Near the wall (y = 0), the functions (v_{x0}, v_{y0}) are not valid solutions of Equations (2.4-13a, b) as ε (= 1/Re) $\longrightarrow 0$.

We need to introduce a new scale Y which is given by

$$Y = \varepsilon^n y \qquad \qquad (2.4\text{-}18)$$

where n is a constant to be determined later.

There is a need to scale v_y and from Equations (2.4-14), we deduce that it has to be scaled as

$$V_y = \varepsilon^n v_y \qquad \qquad (2.4\text{-}19)$$

Equation (2.4-14) becomes

$$\frac{\partial v_x}{\partial x} + \frac{\partial V_y}{\partial Y} = 0 \qquad \qquad (2.4\text{-}20)$$

Combining Equations (2.4-13a, b, 18, 19) yields

$$v_x \frac{\partial v_x}{\partial x} + V_y \frac{\partial v_x}{\partial Y} = -\frac{\partial p}{\partial x} + \varepsilon \left(\frac{\partial^2 v_x}{\partial x^2} + \varepsilon^{2n} \frac{\partial^2 v_x}{\partial Y^2} \right) \tag{2.4-21a}$$

$$\varepsilon^{-n} v_x \frac{\partial V_y}{\partial x} + \varepsilon^{-n} V_y \frac{\partial V_y}{\partial Y} = -\varepsilon^n \frac{\partial p}{\partial Y} + \varepsilon^{1-n} \left(\frac{\partial^2 V_y}{\partial x^2} + \varepsilon^{2n} \frac{\partial^2 V_y}{\partial Y^2} \right) \tag{2.4-21b}$$

To determine n, we require that the highest derivative in Equation (2.2-21a) to be of $O(1)$, that is, $\varepsilon^{1+2n} = O(1)$ which implies

$$n = -1/2 \tag{2.4-22}$$

An alternative way of determining the value of n is by balancing the forces. The left side of Equation (2.4-21a) represents the inertia force and the second term on the right side represents the viscous force. We now require the viscous force to balance the inertia force, that is, $\varepsilon^{1+2n} (\partial^2 v_x / \partial Y^2)$ must be of $O(1)$ which implies Equation (2.4-22).

Note that v_x increases from zero at $y = 0$ to v_∞ in the outer layer, this means that v_x is a fast increasing function of y near the solid wall. This in turn implies that the shear stress τ_{xy} [$= \mu (\partial v_x / \partial y)$] is not negligible and this shear stress provides the required drag. With the value of n given by Equation (2.4-22), Equations (2.4-21a, b) become

$$v_x \frac{\partial v_x}{\partial x} + V_y \frac{\partial v_x}{\partial Y} = -\frac{\partial p}{\partial x} + \frac{\partial^2 v_x}{\partial Y^2} + \varepsilon \frac{\partial^2 v_x}{\partial x^2} \tag{2.4-23a}$$

$$\varepsilon \left(v_x \frac{\partial V_y}{\partial x} + V_y \frac{\partial V_y}{\partial Y} \right) = -\frac{\partial p}{\partial Y} + \varepsilon \left(\varepsilon \frac{\partial^2 V_y}{\partial x^2} + \frac{\partial^2 V_y}{\partial Y^2} \right) \tag{2.4-23b}$$

We expand v_x, V_y, and p as

$$v_x = v_{x0}^{(i)} + \varepsilon v_{x1}^{(i)} + \dots \tag{2.4-24a}$$

$$V_y = V_{y0}^{(i)} + \varepsilon V_{y1}^{(i)} + \dots \tag{2.4-24b}$$

$$p = p_0^{(i)} + \varepsilon p_1^{(i)} + \dots \tag{2.4-24c}$$

Substituting Equations (2.4-24a, b, c) into Equations (2.4-23a, b), we deduce that the zeroth order equations are

$$v_{x0}^{(i)} \frac{\partial v_{x0}^{(i)}}{\partial x} + V_{y0}^{(i)} \frac{\partial v_{x0}^{(i)}}{\partial Y} = -\frac{\partial p_0^{(i)}}{\partial x} + \frac{\partial^2 v_{x0}^{(i)}}{\partial Y^2}$$

(2.4-25a)

$$0 = -\frac{\partial p_0^{(i)}}{\partial Y}$$

(2.4-25b)

Equations (2.4-25a, b) are the boundary-layer equations and the derivation given here is much more precise than that given in Bird et al. (1960).

Equation (2.4-17a, b) are valid in the outer layer (y >> 0) and the solutions do not need to satisfy the condition at y = 0. Equations (2.4-25a, b) are valid in the inner layer and they are solved subject to the boundary conditions at y = 0. Then, we have to match the inner and the outer solutions using Prandtl's matching principle. Equations (2.4-25a) can be solved by the method of similarity variables and are described in Rosenhead (1963) and Chan Man Fong et al. (1997).

Example 2.4-3. **Improving Stokes' solution**

Solve the flow of a Newtonian fluid past a sphere at low Reynolds number.

We choose to work in spherical polar coordinates (r, θ, ϕ) with the origin at the center of a sphere of radius a. It is usual to introduce a stream function $\psi (r, \theta)$ and the velocity components (v_r, v_θ) are given by

$$v_r = -\frac{1}{r^2 \sin \theta} \frac{\partial \psi}{\partial \theta}$$

(2.4-26a)

$$v_\theta = \frac{1}{r \sin \theta} \frac{\partial \psi}{\partial r}$$

(2.4-26b)

For steady flows, the Navier-Stokes equation can be written as (Bird et al., 1960)

$$\frac{1}{r^2 \sin \theta} \left[\frac{\partial \psi}{\partial r} \frac{\partial}{\partial \theta} (E^2 \psi) - \frac{\partial \psi}{\partial \theta} \frac{\partial}{\partial r} (E^2 \psi) - 2 (E^2 \psi) \left(\frac{\partial \psi}{\partial r} \cot \theta - \frac{1}{r} \frac{\partial \psi}{\partial \theta} \right) \right] = \frac{1}{Re} E^4 \psi$$

(2.4-27a)

$$E^2 \equiv \frac{\partial^2}{\partial r^2} + \frac{\sin \theta}{r^2} \frac{\partial}{\partial \theta} \left(\frac{1}{\sin \theta} \frac{\partial}{\partial \theta} \right)$$

(2.4-27b)

$$Re = \rho U a / \mu$$

(2.4-27c)

The boundary conditions are

$$\psi (a, \theta) = \left. \frac{\partial \psi}{\partial r} \right|_{r=a} = 0$$

(2.4-28a,b)

$$\psi(r, \theta) \longrightarrow -\frac{1}{2} r^2 \sin^2 \theta \quad \text{as} \quad r \longrightarrow \infty \tag{2.4-28c}$$

In this example, $\text{Re} \approx 0$ and we set $\text{Re} = \varepsilon$. We expand ψ as

$$\psi = \psi_0 + \varepsilon \psi_1 + \dots \tag{2.4-29}$$

Substituting Equation (2.4-29) into Equation (2.4-27a) and comparing powers of ε yields

$$E^4 \psi_0 = 0 \tag{2.4-30}$$

Note that, in this example, the order of Equation (2.4-27a) is the same as that of Equation (2.4-30) and all the boundary conditions can be satisfied. Equations (2.4-30) is Stokes' equation and ψ_0 is given by (Bird et al., 1960)

$$\psi_0 = -\frac{1}{4}\left(2r^2 - 3r + \frac{1}{r}\right) \sin^2 \theta \tag{2.4-31}$$

The Stokes solution [Equation (2.4-31)] is uniformly valid and we try to improve the solution by computing ψ_1. From Equations (2.4-27a, 28a, b, c), we deduce that ψ_1 satisfies

$$E^4 \psi_1 = N(\psi_0) \tag{2.4-32a}$$

$$\psi_1(a, \theta) = \left.\frac{\partial \psi_1}{\partial r}\right|_{r=a} = 0 \tag{2.4-32b,c}$$

$$[\psi_1(r, \theta)]/r^2 \longrightarrow 0 \quad \text{as} \quad r \longrightarrow 0 \tag{2.4-32d}$$

where N is the operator on the left side of Equation (2.4-27a).

Note that ψ_0 satisfies Equation (2.4-28c) and ψ_1 should be of order less than r^2 as $r \longrightarrow \infty$.

Combining Equations (2.4-31, 32a) yields

$$E^4 \psi_1 = -\frac{9}{4}\left(\frac{2}{r^2} - \frac{3}{r^3} + \frac{1}{r^5}\right) \sin^2 \theta \cos \theta \tag{2.4-33}$$

Equations (2.4-32b, c, 33) suggest that we seek a solution of the form

$$\psi_1 = g(r) \sin^2 \theta \cos \theta \tag{2.4-34}$$

Substituting ψ_1 into Equation (2.4-33), we obtain

$$\frac{d^4 g}{dr^4} - \frac{12}{r^2}\frac{d^2 g}{dr^2} + \frac{24}{r^3}\frac{d g}{dr} = -\frac{9}{4}\left(\frac{2}{r^2} - \frac{3}{r^3} + \frac{1}{r^5}\right) \tag{2.4-35}$$

The complementary function of Equation (2.4-35) is generated by assuming that g is of the form

$$g = r^s \tag{2.4-36}$$

The particular integral is determined by the method of variation of parameters. The solution is found to be

$$\psi_1 = \frac{A}{r^2} + B + Cr^3 + Dr^5 - \frac{3r^2}{16} + \frac{9r}{32} + \frac{3}{32r} \tag{2.4-37}$$

where A–D are arbitrary constants.

The presence of the term $(3r^2/16)$ in the particular integral while there is no r^2 term in the complementary function makes it impossible to satisfy Equation (2.4-32d). The solution ψ_1 is not valid throughout (**Whitehead's paradox**). This non-uniformity is reminiscent of the secular terms in Example 2.1-2. To understand the failure of the regular perturbation method, we compare the magnitude of the inertia force [left side of Equation (2.4-27a)] to the magnitude of the viscous force [right side of Equation (2.4-27a)] as $r \longrightarrow \infty$. The left side of Equation (2.4-27a) is given by the right side of Equation (2.4-33) and as $r \longrightarrow \infty$, the dominant term is $-9/2r^2$ and the inertia force F_i is $O(\varepsilon/r^2)$. We can compute $E^4 \psi_0$ and it is found that the dominant term as $r \longrightarrow \infty$ in $E^4 \psi_0$ is $O(1/r^3)$. The viscous force F_v is $O(1/r^3)$. The ratio

$$\frac{F_i}{F_v} = \varepsilon r = Re\, r \tag{2.4-38a,b}$$

The Stokes approximation in which F_i is neglected is valid only if $\varepsilon r < 1$. However small Re is, εr will be greater than 1 as $r \longrightarrow \infty$. This implies that we need to introduce a new scale for $r \gg 1$, which contracts r. The new scale R is defined by

$$R = r\, Re \tag{2.4-39}$$

By changing the independent variable r to R, Equation (2.4-27a) becomes

$$\frac{1}{R^2 \sin \theta} \left[\frac{\partial \psi}{\partial R} \frac{\partial}{\partial \theta} (D^2 \psi) - \frac{\partial \psi}{\partial \theta} \frac{\partial}{\partial R} (D^2 \psi) - 2 (D^2 \psi) \left(\frac{\partial \psi}{\partial R} \cot \theta - \frac{1}{r} \frac{\partial \psi}{\partial \theta} \right) \right] = \frac{1}{Re^2} D^4 \psi \tag{2.4-40a}$$

$$D^2 = \frac{\partial^2}{\partial R^2} + \frac{\sin \theta}{R^2} \frac{\partial}{\partial \theta} \left(\frac{1}{\sin \theta} \frac{\partial}{\partial \theta} \right) \tag{2.4-40b}$$

The Stokes solution represents the inner solution and is valid near the sphere. The outer solution (Oseen solution) is obtained by solving Equation (2.4-40a) subject to the condition given by Equation (2.4-32d). Rather than solving Equation (2.4-40a), it has been found that it is simpler to guess the form of the Oseen solution by applying the matching principle. Details of the calculations are given in Nayfeh (1973) and Van Dyke (1975).

●

Prandtl's matching principle works for the first term but might not work for higher order terms. Van Dyke has proposed another method of matching the inner and outer solutions. We illustrate Van Dyke's matching principle in the next example. We shall also discuss the problem of locating the boundary layer.

Example 2.4-4. Van Dyke's matching principle

Solve the boundary-layer problem

$$\varepsilon \frac{d^2 y}{dx^2} + \frac{dy}{dx} = \frac{1}{2} x^2 \qquad (2.4\text{-}41a)$$

$$y(0) = 1, \qquad y(1) = 1/6 \qquad (2.4\text{-}41b,c)$$

by the method of matched asymptotic expansions.

Equation (2.4-41a) is a linear non-homogeneous equation and it can be solved exactly. The exact solution y_e is

$$y_e = \frac{x^3}{6} - \frac{\varepsilon x^2}{2} + \varepsilon^2 x + \frac{e^{-x/\varepsilon} - e^{-1/\varepsilon} + (\varepsilon/2 - \varepsilon^2)(1 - e^{-x/\varepsilon})}{(1 - e^{-1/\varepsilon})} \qquad (2.4\text{-}42)$$

We assume that the boundary layer is near $x = 1$. We introduce the scale X_1 defined by

$$X_1 = (1 - x)/\varepsilon \qquad (2.4\text{-}43)$$

We denote the dependent variable by $Y(X_1)$ and, in terms of the new variables, Equation (2.4-41a) becomes

$$\frac{d^2 Y}{dX_1^2} - \frac{dY}{dX_1} = \varepsilon(1 - \varepsilon X_1)^2 \qquad (2.4\text{-}44)$$

The appropriate boundary condition is at $x = 1$ and is written as

$$Y_1(0) = 1/6 \qquad (2.4\text{-}45)$$

We expand Y as

$$Y = Y_0^{(i)} + \varepsilon Y_1^{(i)} + \dots \tag{2.4-46}$$

The function $Y_0^{(i)}$ satisfies

$$\frac{d^2 Y_0^{(i)}}{dX_1^2} - \frac{dY_0^{(i)}}{dX_1} = 0 \tag{2.4-47}$$

The solution that satisfies Equation (2.4-45) is

$$Y_0^{(i)} = A_0 (1 - e^{X_1}) + \frac{1}{6} e^{X_1} \tag{2.4-48}$$

The outer region is near $x = 0$, the independent variable is x, and the boundary condition is given by Equation (2.4-41b).

The function $y(x)$ is expanded as

$$y = y_0^{(0)} + \varepsilon y_1^{(0)} + \dots \tag{2.4-49}$$

Substituting Equation (2.4-49) into Equation (2.4-41a) and comparing powers of ε yields

$$\frac{dy_0^{(0)}}{dx} = \frac{1}{2} x^2 \tag{2.4-50}$$

The solution is

$$y_0^{(0)} = 1 + x^3/6 \tag{2.4-51}$$

To determine A_0 we apply Prandtl's matching principle, that is

$$\lim_{X_1 \to \infty} Y_0^{(i)} = \lim_{x \to 1} y_0^{(0)} \tag{2.4-52}$$

The left side tends to infinity but the right side tends to 7/6 and the matching fails. This suggests that the boundary layer is not chosen properly. The boundary layer is near $x = 0$. This can be seen from the exact solution y_e. On letting $\varepsilon \to 0$, $y_e \to x^3/6$ and on setting $x = 0$, we obtain zero and not unity as required by the boundary condition. If we first let $x = 0$, $y_e(0)$ is unity, that is to say

$$\lim_{\substack{x \to 0 \\ \varepsilon \to 0}} y_e \neq \lim_{\substack{\varepsilon \to 0 \\ x \to 0}} y_e$$

The independent variable in the boundary layer near $x = 0$ is denoted by X and is defined by

$$X = x/\varepsilon \tag{2.4-53}$$

The dependent variable is denoted by $Y(X)$ and Equation (2.4-41a) becomes

$$\frac{d^2Y}{dX^2} + \frac{dY}{dX} = \frac{1}{2}\,\varepsilon^3 X^2 \qquad\qquad (2.4\text{-}54)$$

The appropriate boundary condition is

$$Y(0) = 1 \qquad\qquad (2.4\text{-}55)$$

Expanding Y as in Equation (2.4-46), we obtain

$$Y_0^{(i)} = A_0\,(1 - e^{-X}) + e^{-X} \qquad\qquad (2.4\text{-}56)$$

The outer solution is generated in the usual manner, the boundary condition associated with the outer solution is given by Equation (2.4-41c). The solution $y_0^{(0)}$ is

$$y_0^{(0)} = x^3/6 \qquad\qquad (2.4\text{-}57)$$

Applying Prandtl's matching principle yields

$$\lim_{X \to \infty} A_0\,(1 - e^{-X}) + e^{-X} = \lim_{x \to 0} x^3/6 \quad\Rightarrow\quad A_0 = 0 \qquad\qquad (2.4\text{-}58)$$

The composite solution $y_0^{(c)}$ is

$$y_0^{(c)} = Y_0^{(i)} + y_0^{(0)} - y_0^{(0L)} = e^{-x/\varepsilon} + x^3/6 \qquad\qquad (2.4\text{-}59\text{a,b})$$

Comparing Equations (2.4-42, 59b), we deduce that $y_0^{(c)}$ is correct to $O(\varepsilon^0)$.

To order ε, we have

$$\frac{d^2Y_1^{(i)}}{dX^2} + \frac{dY_1^{(i)}}{dX} = 0 \qquad\qquad (2.4\text{-}60\text{a})$$

$$Y_1^{(i)}(0) = 0 \qquad\qquad (2.4\text{-}60\text{b})$$

$$\frac{dy_1^{(0)}}{dx} = -\frac{d^2y_0^{(0)}}{dx^2} = -x \qquad\qquad (2.4\text{-}61\text{a,b})$$

$$y_1^{(0)}(1) = 0 \qquad\qquad (2.4\text{-}61\text{c})$$

The solutions are

$$Y_1^{(i)} = A_1\,(1 - e^{-X}) \qquad\qquad (2.4\text{-}62)$$

$$y_1^{(0)} = (1 - x^2)/2 \qquad\qquad\qquad (2.4\text{-}63)$$

where A_1 is an arbitrary constant.

We now introduce Van Dyke's matching principle to determine the constants A_0 and A_1. The inner solution $[Y_0^{(i)}(X) + \varepsilon Y_1^{(i)}(X) + ...]$ is rewritten in terms of the outer independent variable x and then expanded in powers of ε. In this example, we write

$$Y = Y_0^{(i)}(X) + \varepsilon Y_1^{(i)}(X) + ... \qquad\qquad (2.4\text{-}64a)$$

$$= A_0(1 - e^{-X}) + e^{-X} + \varepsilon A_1(1 - e^{-X}) + ... \qquad\qquad (2.4\text{-}64b)$$

$$= A_0(1 - e^{-x/\varepsilon}) + e^{-x/\varepsilon} + \varepsilon A_1(1 - e^{-x/\varepsilon}) + ... \qquad\qquad (2.4\text{-}64c)$$

$$\approx A_0 + \varepsilon A_1 + ... \qquad\qquad (2.4\text{-}64d)$$

Likewise, the outer solution $[y_0^{(0)}(x) + \varepsilon y_1^{(0)}(x) + ...]$ is expressed in terms of X and expanded in powers of ε. In the present example, y is written as

$$y = y_0^{(0)}(x) + \varepsilon y_1^{(0)}(x) + ... \qquad\qquad (2.4\text{-}65a)$$

$$= x^3/6 + \varepsilon(1 - x^2)/2 \qquad\qquad (2.4\text{-}65b)$$

$$\approx \varepsilon^3 X^3/6 + \varepsilon/2 - \varepsilon^3 X^3/2 \qquad\qquad (2.4\text{-}65c)$$

We now match the two expansions, that is, we compare powers of ε in Equation (2.4-64d) to those in Equation (2.4-65c). We deduce

$$A_0 = 0, \qquad A_1 = 1/2 \qquad\qquad (2.4\text{-}66a,b)$$

Combining Equations (2.4-56, 57, 62, 63, 66a, b) yields

$$y_c = x^3/6 + \varepsilon(1 - x^2)/2 + e^{-x/\varepsilon} + \varepsilon(1 - e^{-x/\varepsilon})/2 - \varepsilon/2 + O(\varepsilon^2) \qquad (2.4\text{-}67a)$$

$$= x^3/6 - \varepsilon x^2/2 + e^{-x/\varepsilon}(1 - \varepsilon) + \varepsilon/2 + O(\varepsilon^2) \qquad\qquad (2.4\text{-}67b)$$

From Equations (2.4-66a, b), we deduce that in the overlap region, the value of the dependent variable is $\varepsilon/2$. Comparing Equations (2.4-42, 67b), we conclude that y_c is correct to the order indicated.

●

Example 2.4-5. **Slider bearing**

Solve Reynolds' equation for a slider bearing.

This problem was solved by Di Prima (1968, 1969) using the method of matched asymptotic expansions.

We consider an infinitely long slider bearing as shown in Figure 2.4-1. The Reynolds equation for the pressure distribution $p(x)$ in an isothermal, compressible gas film, at steady state can be written in dimensionless form (Cameron, 1971) as

$$\frac{d}{dx}\left(h^3 p \frac{dp}{dx}\right) = \Lambda \frac{d}{dx}(ph) \tag{2.4-68}$$

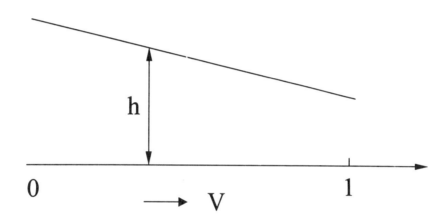

FIGURE 2.4-1   Slider bearing

The spatial variable x, the film thickness h, and the pressure p have been made dimensionless with respect to B (the length of the bearing in the flow direction), C (the clearance at $x = 1$), and p_a (the ambient pressure) respectively. The bearing number Λ is defined by

$$\Lambda = 6BV\mu/p_a C^2 \tag{2.4-69}$$

where μ is the viscosity and V is the velocity of the lower surface, as shown in Figure 2.4-1.

The boundary conditions are

$$p(0) = 1, \qquad p(1) = 1 \tag{2.4-70a,b}$$

We seek a solution for $\Lambda \longrightarrow \infty$ and we write $\varepsilon = 1/\Lambda$. Equation (2.4-68) can be written as

$$\varepsilon \frac{d}{dx}\left(h^3 p \frac{dp}{dx}\right) = \frac{d}{dx}(ph) \tag{2.4-71}$$

Note that the coefficient of the highest derivative (d^2p/dx^2) is ε and we can expect the existence of a boundary layer. We seek the outer solution by expanding p as

$$p = p_0^{(0)} + \varepsilon p_1^{(0)} + \dots \tag{2.4-72}$$

We assume that the outer layer is in the neighborhood of $x = 0$, since the boundary layer (the region where one expects rapid changes of p) should be near $x = 1$.

Substituting Equation (2.4-72) into Equation (2.4-71) yields to various powers of ε

$$\varepsilon^0: \quad \frac{d}{dx}(p_0^{(0)} h) = 0 \tag{2.4-73a}$$

$$\varepsilon: \quad \frac{d}{dx}(p_1^{(0)} h) = \frac{d}{dx}\left(h^3 p_0^{(0)} \frac{dp_0^{(0)}}{dx}\right) \tag{2.4-73b}$$

The solution of Equation (2.4-73a) subject to Equation (2.4-70a) is

$$p_0^{(0)} = h(0)/h(x) \tag{2.4-74}$$

Substituting $p_0^{(0)}$ into Equation (2.4-73b) and integrating yields

$$p_1^{(0)} h = -h^2(0) h'(x) + A_1 \tag{2.4-75}$$

where A_1 is a constant, $h'(x)$ is dh/dx.

Imposing the boundary condition $[p_1^{(0)}(0) = 0]$ results in

$$p_1^{(0)} = h^2(0)[h'(0) - h'(x)]/h \tag{2.4-76}$$

Note that $p_0^{(0)}$ does not satisfy the boundary condition $x = 1$. To obtain the inner solution, we introduce a new scale X defined by

$$X = \varepsilon^n (1 - x) \tag{2.4-77}$$

We write $p(x)$ as $P(X)$, $h(x)$ as $H(X)$, and Equation (2.4-71) becomes

$$\varepsilon^{2n+1} \frac{d}{dX}\left(H^3 P \frac{dP}{dX}\right) = -\varepsilon^n \frac{d}{dX}(PH) \tag{2.4-78}$$

Requiring both sides of the equation to be of the same order of magnitude implies that

$$n = -1 \tag{2.4-79}$$

X is now given by

$$X = (1-x)/\varepsilon \tag{2.4-80}$$

Here, we consider the case of a wedge slider with $h(x)$ given by

$$h(x) = 1 + \alpha(1-x) \tag{2.4-81}$$

where α is a constant.

From Equations (2.4-80, 81), we deduce that $H(x)$ is given by

$$H(x) = 1 + \alpha\varepsilon X \tag{2.4-82}$$

We expand P as

$$P = P_0^{(i)} + \varepsilon P_1^{(i)} + \dots \tag{2.4-83}$$

Combining Equations (2.4-78, 79, 82, 83) and comparing powers of ε yields

$$\varepsilon^0: \quad \frac{d}{dX}\left(P_0^{(i)}\frac{dP_0^{(i)}}{dX}\right) = -\frac{dP_0^{(i)}}{dX} \tag{2.4-84a}$$

$$\varepsilon: \quad \frac{d}{dX}\left(P_0^{(i)}\frac{dP_1^{(i)}}{dX}\right) = -\frac{d}{dX}\left(P_1^{(i)} + \alpha X P_0^{(i)}\right) - \frac{d}{dX}\left(3\alpha X P_0^{(i)}\frac{dP_0^{(i)}}{dX} + P_1^{(i)}\frac{dP_0^{(i)}}{dX}\right) \tag{2.4-84b}$$

The appropriate boundary condition is at $X = 0$ $(x = 1)$ and is

$$P(0) = 1 \tag{2.4-85}$$

This implies that

$$P_0^{(i)}(0) = 1, \quad P_1^{(i)}(0) = 0, \quad \dots \tag{2.4-86a,b}$$

Integrating Equation (2.4-84a) yields

$$P_0^{(i)}\frac{dP_0^{(i)}}{dX} = -P_0^{(i)} + A_0 \tag{2.4-87}$$

where A_0 is a constant.

Equation (2.4-87) can also be written as

$$\frac{P_0^{(i)} \, dP_0^{(i)}}{P_0^{(i)} - A_0} = - dX \tag{2.4-88a}$$

or $\qquad \left(1 + \dfrac{A_0}{P_0^{(i)} - A_0}\right) dP_0^{(i)} = - dX \tag{2.4-88b}$

Integrating Equation (2.4-88b) yields

$$P_0^{(i)} + A_0 \ell n \, (P_0^{(i)} - A_0) = - X + B_0 \tag{2.4-89}$$

where B_0 is an arbitrary constant.

Applying the boundary condition [Equation (2.4-86a)] results in

$$B_0 = 1 + A_0 \ell n \, (1 - A_0) \tag{2.4-90}$$

The solution can be written as

$$P_0^{(i)} + A_0 \ell n \, [(P_0^{(i)} - A_0)/(1 - A_0)] = 1 - X \tag{2.4-91a}$$

or $\qquad [(P_0^{(i)} - A_0)/(1 - A_0)] = \exp\,[(1 - X - P_0^{(i)})/A_0] \tag{2.4-91b}$

The constant A_0 is determined by matching $P_0^{(i)}$ to $p_0^{(0)}$ and we adopt Prandtl's matching principle which can be written as

$$\lim_{X \to \infty} P_0^{(i)} = \lim_{x \to 1} p_0^{(0)} = (1 + \alpha) \tag{2.4-92a,b}$$

Equation (2.4-92b) follows from Equations (2.4-74, 81).

As $X \longrightarrow \infty$, the right side of Equation (2.4-91b) tends to zero and $P_0^{(i)} \longrightarrow (1 + \alpha)$ and this implies that

$$A_0 = 1 + \alpha \tag{2.4-93}$$

Combining Equations (2.4-91a, 93) yields

$$P_0^{(i)} + (1 + \alpha)\, \ell n \, [(1 + \alpha - P_0^{(i)})/\alpha] = 1 - X \tag{2.4-94}$$

The composite solution $p_0^{(c)}$ is given by

$$p_0^{(c)} = p_0^{(0)} + P_0^{(i)} - p_0^{(0L)} \tag{2.4-95a}$$

$$= (1 + \alpha) / [1 + \alpha (1 - x)] + P_0^{(i)} - (1 + \alpha) \qquad (2.4\text{-}95b)$$

where $P_0^{(i)}$ is given by Equation (2.4-94) in an implicit form.

Higher order approximations can be computed and are given in Di Prima (1969).

●

In this chapter, we have shown that the regular perturbation method may fail (a) if the region is infinite ($|x| \longrightarrow \infty$) and εx may be greater than one, (b) if ε is the coefficient of the highest derivative in the differential equation.

Van Dyke (1975) has stated a necessary physical criterion for uniformity. According to this criterion, a perturbation solution is uniformly valid unless ε is the ratio of two lengths or two times. In Example 2.4-5, the bearing number Λ [Equation (2.4-69)] involves two lengths B and C and when $\Lambda \longrightarrow \infty$ [$\varepsilon (= 1/\Lambda) \longrightarrow 0$], either $B \longrightarrow \infty$ or $C \longrightarrow 0$. In this case, we can expect the possibility of non-uniformity.

To render the solution uniformly valid, we have to introduce new scales. Several techniques have been proposed to solve the singular problems and we have described three of them, namely the multiple scales, the method of averaging, and the method of matched asymptotic expansions. Morrison (1966) has compared the method of averaging with the method of multiple scales. Wollkind (1977) made a comparison between the solutions generated by the technique of matched asymptotic expansions with that of multiple scales.

In Section 4, we have discussed the existence of a boundary layer where the dependent variable undergoes rapid changes. By introducing a new scale, the equation is simplified without losing its main properties. Even if the boundary-layer equation cannot be solved analytically, it is easier to solve the boundary layer numerically rather than solving the full equation numerically. Wilcox (1995) has discussed the effectiveness of combining numerics and asymptotics.

In the next chapter, we shall discuss the concept of stability which was introduced in Chapter 1.

PROBLEMS

1. A particle of unit mass is dropped from rest at a height h. The air resistance is assumed to be proportional to the square of the velocity. The equation of motion can be written as

$$\frac{d^2 y}{dt^2} = -\varepsilon \left(\frac{dy}{dt} \right)^2 + g$$

Assume ε to be small and use the method of perturbation to determine the time at which the particle hits the ground.

2. Solve the equation

$$\frac{d^2y}{dt^2} + y - \varepsilon y^2 = 0$$

subject to the following initial conditions

$$y(0) = 1, \qquad \left.\frac{dy}{dt}\right|_{t=0} = 0$$

by the method of perturbation. The parameter ε is small.

3. The equations governing the sedimentation of a charged particle can be written in dimensionless form as

$$\nabla^2 \underline{v} = \underline{\nabla} P - \rho \underline{g} + \beta \zeta_a \underline{\nabla} \Psi \nabla^2 \Psi$$

$$\underline{\nabla} \cdot \underline{v} = 0$$

$$\zeta_a \nabla^2 \Psi = -\frac{1}{2}(\kappa a)^2 (C_+ - C_-)$$

$$\underline{v} \cdot \underline{\nabla} C_+ = \underline{\nabla}\,[\underline{\nabla} C_+ + \zeta_a C_+ \underline{\nabla} \Psi]$$

$$\underline{v} \cdot \underline{\nabla} C_- = q\underline{\nabla}\,[\underline{\nabla} C_- - \zeta_a C_- \underline{\nabla} \Psi]$$

where $\underline{v}$, P, $\underline{g}$, Ψ, and $C_\pm$ are dimensionless form of the velocity field, hydrostatic pressure, gravity, local electrical potential, and ion concentration respectively. The parameters β, ζ_a, κa, and q are the dimensionless electrical force constant, surface potential, inverse double layer thickness, and ratio of ion diffusivities respectively.

Pujar and Zydney (1996) sought a solution in powers of a Peclet number Pe which is defined as

$$Pe = U_a/D_+$$

where U is the sedimentation velocity, a is the characteristic particle dimension, and D_+ is the positive ion diffusivity.

Noting that when the Peclet number is zero, the system is in equilibrium, expand $\underline{v}$, C, Ψ, and P appropriately. Obtain the zeroth and the first order equations.

4. Solve the system

$$\frac{\partial^2 u}{\partial x^2} + \varepsilon \frac{\partial^2 u}{\partial y^2} = 0, \qquad 0 < x < 1, \ 0 < y < 1, \ \varepsilon \ll 1$$

$$u(x, 0) = u(x, 1) = u(1, y) = 0, \qquad u(0, y) = \sin \pi y$$

by the method of perturbation. Is the solution valid if the region is $0 < x < \infty$?

5. Obtain the solution of

$$\frac{\partial^2 u}{\partial r^2} + \frac{1}{r} \frac{\partial u}{\partial r} + \frac{1}{r^2} \frac{\partial^2 u}{\partial \theta^2} = 0, \qquad 0 < r < 1 + \varepsilon \cos\theta, \ 0 < \theta < 2\pi, \ \varepsilon \ll 1$$

subject to

$$u \text{ is finite at the origin,} \qquad u(1 + \varepsilon \cos \theta, \theta) = \cos \theta$$

by the method of perturbation. Note that the boundary is a distorted unit circle.

6. By introducing two time scales $(T_0 = t, \ T_1 = \varepsilon t)$, show that the solution of Rayleigh's equation

$$\ddot{x} - \varepsilon \dot{x} (1 - \dot{x}^2/3) + x = 0$$

subject to

$$x(0) = 1, \qquad \dot{x}(0) = 0$$

is

$$x(t) = 2[1 + 3e^{-\varepsilon t}]^{-1/2} \cos t + \dots$$

7. Van Dyke (1987) has considered the potential problem in a hyperbolic strip. The strip is bounded by the hyperbolas

$$y = \pm \sqrt{(1 + \varepsilon^2 x^2)}$$

If ψ satisfies Laplace's equation in the strip and assumes the values ± 1 on the upper and lower edges, determine ψ by the method of perturbation. To eliminate the appearance of ε in the boundary replace εx by X.

8. In Example 2.2-3, we have obtained the solution of Equation (2.2-33) near $\omega = 2$. Obtain the solution near $\omega = 1$.

9.   Use the Lindstedt-Poincaré method to solve the equation

$$\frac{d^2x}{dt^2} + x + \varepsilon x^2 = 0$$

The initial conditions are

$$x(0) = 1, \qquad \frac{dx}{dt}\bigg|_{t=0} = 0$$

Verify that to $O(\varepsilon)$ the solution is regular and obtain the solution to $O(\varepsilon^2)$.

10.   Solve the equation

$$\frac{d^2x}{dt^2} + x = \varepsilon x^3$$

subject to

$$x(0) = 1, \qquad \frac{dx}{dt}\bigg|_{t=0} = 0$$

by the Krylov-Bogoliubov and the KBM method.

11.   Hanks (1971) has deduced that for a one-dimensional non-dissipative steady flow, the heat transfer equation is

$$\varepsilon \frac{d^2T}{dx^2} + x \frac{dT}{dx} - xT = 0, \qquad 0 \le x \le \ell, \ \varepsilon \ll 1$$

$$T(0) = T_0, \qquad T(\ell) = T_1$$

Use the method of matched asymptotic expansions to obtain an approximative solution.

12.   The equation describing the motion of a heavily damped string is

$$\varepsilon \left(\frac{\partial^2 u}{\partial t^2} - \frac{\partial^2 u}{\partial x^2} \right) + \frac{\partial u}{\partial t} = 0, \qquad t > 0, \ -\infty < x < \infty, \ \varepsilon \ll 1$$

The initial conditions are

$$u(x, 0) = f(x), \qquad \frac{\partial u}{\partial t}\bigg|_{t=0} = g(x)$$

Show that the regular perturbation method does not provide a solution satisfying the two initial conditions. Use the method of matched asymptotic expansions to obtain a uniformly valid solution. Near $t \approx 0$, introduce a new variable τ defined by

$$\tau = t/\varepsilon$$

CHAPTER 3
STABILITY

3.1 INTRODUCTION

Stability commonly means durability, constancy, steadiness, immobility, ... In science, we need a precise definition of stability and this has led to the existence of several types of stability. Stability was first introduced in mechanics to describe the position of equilibrium of a particle. Suppose that a particle is maintained in equilibrium at a position $\underline{x}_e$ under a set of forces. Next, its position and velocity are slightly perturbed. The point $\underline{x}_e$ is a **stable equilibrium point** if the particle remains in the neighborhood of $\underline{x}_e$ at all subsequent times. The point $\underline{x}_e$ is **asymptotically stable** if the particle eventually returns to $\underline{x}_e$. This type of stability was briefly discussed in Chapter 1.

Other definitions of stability have been proposed. In celestial mechanics, Lagrange stated that the trajectory of a planet is stable if its major axis remains bounded even after it has been perturbed. In this definition, stability is associated with boundedness. According to Poisson, a trajectory is stable if it passes arbitrarily close to every point of its past trajectory. Thus within the same subject, it is possible to have more than one definition of stability.

The concept of stability is associated with various branches of science, such as mechanics, control theory, chemical kinetics, economics, etc., and this resulted in several definitions of stability. However, they are all related to the definitions given in the first paragraph. In all cases, one considers an unperturbed system, whose stability is to be investigated. A perturbation is applied to the system and the states of the perturbed system can be characterized by a norm and the change in the norm can be used to define stability. There is more than one definition of the norm. In the next section, we define various types of norms and associated stabilities.

3.2 DEFINITIONS

We assume that the system under consideration can be described by a set of equations, which can be written as

$$\frac{d\underline{x}}{dt} = \underline{f}(\underline{x}, t) \tag{3.2-1}$$

where $\underline{x}$ and $\underline{f}$ are column vectors with n elements each.

The independent variable t usually denotes time. We also assume that $\underline{f}$ is smooth enough to ensure that Equation (3.2-1) has a unique solution that depends on the initial conditions. To examine the stability of the solutions of Equation (3.2-1), we need to introduce a measure of distance (norm) between solutions at particular times. Several measures of the size of $\underline{x}$, **norms** of $\underline{x}$, have been defined. The most common one is the **Euclidean norm** which is defined by

$$\| \underline{x} \| = \left(\sum_{i=1}^{n} | x_i |^2 \right)^{1/2} \tag{3.2-2a}$$

Norms can also be defined as

$$\| \underline{x} \| = \sum_{i=1}^{n} | x_i | \tag{3.2-2b}$$

$$\| \underline{x} \| = \max \left\{ | x_1 |, | x_2 |, \ldots , | x_n | \right\} \tag{3.2-2c}$$

We next define **Liapunov stability**.

The solution $\underline{x}^*$ of Equation (3.2-1) is stable in the sense of Liapunov at time t_0 if there exists a $\delta (\varepsilon, t_0)$ such that

$$\| \underline{x} (t_0) - \underline{x}^*(t_0) \| < \delta \quad \Rightarrow \quad \| \underline{x} (t) - \underline{x}^*(t) \| < \varepsilon \tag{3.2-3}$$

for all $t > t_0$ and $\underline{x} (t)$ is any other solution.

Statement (3.2-3) means that if two solutions are close at t_0, they are close at all times $t > t_0$. It can be proved (Cesari, 1971) that if Statement (3.2-3) is true for t_0, it is true for any $t_1 > t_0$. If Statement (3.2-3) is not true, $\underline{x}^*(t)$ is **unstable in the Liapunov sense**.

If δ is independent of t_0, $\underline{x}^*$ is **uniformly stable**.

The solution $\underline{x}^*$ is **quasi-asymptotically stable** if

$$\| \underline{x} (t_0) - \underline{x}^*(t_0) \| < \delta \quad \Rightarrow \quad \lim_{t \longrightarrow \infty} \| \underline{x} (t) - \underline{x}^*(t) \| = 0 \tag{3.2-4}$$

If $\underline{x}^*$ is both stable and quasi-asymptotically stable, it is **asymptotically stable**.

We illustrate the definitions of Liapunov stability in the next example.

Example 3.2.1. **Determination of stability**

Examine the stability of the following systems

(i) $\dfrac{dx}{dt} = -xt$ (3.2-5a)

(ii) $\dfrac{dx}{dt} = xt$ (3.2-5b)

(iii) $\dfrac{dx_1}{dt} = -x_2, \qquad \dfrac{dx_2}{dt} = x_1$ (3.2-5c,d)

Without loss of generality we may assume $t_0 = 0$.

(i)   One solution of Equation (3.2-5a) is

$$x(t) = x_0 \exp(-t^2/2)$$ (3.2-6)

where x_0 is the value of x at $t = 0$.

Another solution of Equation (3.2-5a) is

$$x^*(t) = x_0^* \exp(-t^2/2)$$ (3.2-7)

where x_0^* is $x^*(0)$.

We assume that

$$|x_0 - x_0^*| < \delta$$ (3.2-8)

From Equations (3.2-6, 7), we deduce that

$$|x(t) - x^*(t)| = |x_0 - x_0^*| \exp(-t^2/2)$$ (3.2-9a)

$$< \delta \exp(-t^2/2)$$ (3.2-9b)

Inequality (3.2-9b) implies that the system is stable [Inequality (3.2-3) is satisfied with $\delta = \varepsilon$] and further, $|x(t) - x^*(t)|$ tends to zero as t tends to infinity. The system described by Equation (3.2-5a) is asymptotically stable.

(ii)   Two possible solutions of Equation (3.2-5b) are

$$x(t) = x_0 \exp(t^2/2)$$ (3.2-10a)

$$x^*(t) = x_0^* \exp(t^2/2)$$ (3.2-10b)

Suppose there is a δ such that

$$|x_0 - x_0^*| < \delta$$ (3.2-11)

From Equations (3.2-10a, b), we deduce

$$|x(t) - x^*(t)| = |x_0 - x_0^*| \exp(t^2/2) \qquad (3.2\text{-}12)$$

It is seen from Equation (3.2-12) that however large we choose ε to be, we can choose a t such that $|x(t) - x^*(t)|$ is larger than ε. We conclude that the system is unstable.

(iii)   The two possible solutions of Equations (3.2-5c, d) are

$$x_1(t) = x_{10}\cos t - x_{20} \sin t \qquad (3.2\text{-}13a)$$

$$x_2(t) = x_{20}\cos t + x_{10} \sin t \qquad (3.2\text{-}13b)$$

and

$$x_1^*(t) = x_{10}^*\cos t - x_{20}^* \sin t \qquad (3.2\text{-}14a)$$

$$x_2^*(t) = x_{20}^*\cos t + x_{10}^* \sin t \qquad (3.2\text{-}14b)$$

where x_{i0} and x_{i0}^* $(i = 1, 2)$ are $x_i(0)$ and $x_i^*(0)$.

Using the Euclidean norm, we obtain

$$\| \underline{x}(0) - \underline{x}^*(0) \| = \sqrt{(x_{10} - x_{10}^*)^2 + (x_{20} - x_{20}^*)^2} \qquad (3.2\text{-}15a)$$

$$\| \underline{x}(t) - \underline{x}^*(t) \| = \sqrt{[x_1(t) - x_1^*(t)]^2 + [x_2(t) - x_2^*(t)]^2} \qquad (3.2\text{-}15b)$$

$$= \sqrt{(x_{10} - x_{10}^*)^2 + (x_{20} - x_{20}^*)^2} \qquad (3.2\text{-}15c)$$

We deduce that the system is stable but not quasi-asymptotically stable.

●

It is sometimes convenient to introduce a new variable $\underline{z}$ defined by

$$\underline{z} = \underline{x} - \underline{x}^* \qquad (3.2\text{-}16)$$

Equation (3.2-1) now becomes

$$\frac{d\underline{z}}{dt} = \underline{f}(\underline{x}^* + \underline{z}, t) - \underline{f}(\underline{x}^*, t) \qquad (3.2\text{-}17a)$$

$$= \underline{g}(\underline{z}, t) \qquad (3.2\text{-}17b)$$

where $\underline{g}\,(\underline{0}, t) = 0$ for all t.

The origin is an equilibrium point of Equation (3.2-17b) and the stability of solution $\underline{x}^*$ of Equation (3.2-1) is equivalent to the stability of the equilibrium point $(\underline{z} = \underline{0})$ of Equation (3.2-17b). In Chapter 1, we have discussed the stability of equilibrium points of linear and almost linear systems. In this chapter, we shall consider non-linear systems.

If all the solutions of Equation (3.2-1 or 3.2-17b) are bounded for all $t \ge 0$, the solution is **stable in the Lagrangian sense**.

This is expressed mathematically as

$$\parallel \underline{x}\,(t; \underline{x}_0, t_0) - \underline{x}^*(t; \underline{x}_0^*, t_0) \parallel \;\le\; B\,(t_0) \tag{3.2-18}$$

for all $t \ge t_0$.

If B is independent of t_0, then the solution is **uniformly bounded**.

If Equation (3.2-1) has a periodic solution, the trajectory is a closed curve C. It is natural to consider the periodic solution to be stable if a second curve starting near the periodic orbit stays near this trajectory at all subsequent times. If $\underline{x}^*$ is the periodic solution and $\underline{x}$ is another solution, then the periodic solution is stable if Statement (3.2-3) is true. But this definition is too restrictive and has not taken into account the possibility of a phase difference. For simplicity, suppose the orbits of the two solutions $\underline{x}$ and $\underline{x}^*$ are two concentric circles C_1 and C_2 as shown in Figure 3.2-1. The two solutions have different periods leading to a phase difference. Consider two points, P_0 and Q_0, at time t_0 as shown in Figure 3.2-1. At time t_1, the two points are at diametrically opposite positions P_1 and Q_1 as illustrated in Figure 3.2-1. These two points are no longer close, yet C_1 and C_2 are close and we need to redefine the notion of stability in such a way that the solution $\underline{x}^*$ is considered stable. We can do this by eliminating the parameter t and by defining a distance independent of time.

We define the distance of a point $\underline{x}_1$ to a periodic orbit C by

$$\text{dist}\,(\underline{x}_1, C) \;=\; \min\,\{ \parallel \underline{x}_1 - \underline{x} \parallel,\; \text{for all } \underline{x} \in C \} \tag{3.2-19}$$

The periodic solution $\underline{x}^*$, whose closed orbit is C, is **orbitally stable** if for any $\varepsilon > 0$, there is a $\delta > 0$ such that

$$\text{dist}\,(\underline{x}\,(t_0), C) < \delta \quad \Rightarrow \quad \text{dist}\,(\underline{x}\,(t), C) < \varepsilon \;\text{ for all } t > t_0 \tag{3.2-20}$$

where $\underline{x}\,(t)$ is any solution of Equation (3.2-1).

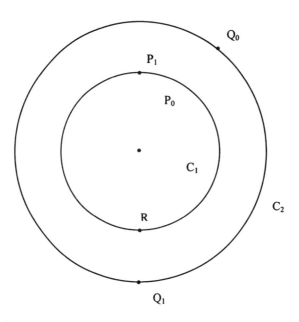

FIGURE 3.2-1   Orbital stability

According to the definition given by Equation (3.2-19), dist (Q_1, C_1) in Figure 3.2-1 is Q_1R which is equal to Q_0P_0 and the orbit C_1 is orbitally stable.

A periodic orbit C is **orbitally asymptotically stable** if it is orbitally stable and if

$$\text{dist}\left(\underline{x}(t_0), C\right) < \delta \quad \Rightarrow \quad \text{dist}\left(\underline{x}(t), C\right) \longrightarrow 0 \text{ as } t \longrightarrow \infty \qquad (3.2\text{-}21)$$

An example of an orbitally asymptotically stable orbit is shown in Figure 1.3-1.

The concept of orbital stability can be extended to orbits which are not closed, that is, the solutions are not periodic. The definitions given by Equations (3.2-20, 21) still apply with C being a non-closed orbit. According to this definition, if two orbits C_1 and C_2 corresponding to the solutions $\underline{x}$ and $\underline{x}^*$ are close to each other at time $t = t_0$ and they remain close at all subsequent times $t > t_0$, $\underline{x}^*$ is orbitally stable. If C_1 and C_2 converge to each other, $\underline{x}^*$ is asymptotically orbitally stable.

In fluid mechanics, the stability of a steady flow described by a velocity field $\underline{v}(\underline{x})$ is investigated by superimposing a disturbance $\underline{u}(\underline{x}, t)$ on $\underline{v}(\underline{x})$. If $\underline{u}(\underline{x}, t) \longrightarrow 0$ as $t \longrightarrow \infty$, the flow is stable, if $\underline{u}(\underline{x}, t)$ is bounded, it is neutrally stable, otherwise it is unstable. In viscous flow, the parameter controlling the stability of the flow is the Reynolds number Re. If Re is small, the flow is stable and there is a critical Reynolds number Re_c at which the flow becomes unstable. The steady solution $\underline{v}(\underline{x})$ bifurcates at Re_c (see Chapter 1). If the flow is unsteady, the definition of stability given earlier is not appropriate. The growth or decay of $\underline{u}(\underline{x}, t)$ can be due to the fluctuation in the basic unsteady flow. Other definitions need to be introduced. Drazin and Reid (1979) have discussed this case. In Section 3.5, we shall give an example of hydrodynamic stability.

So far we have assumed that $\underline{f}$ in Equation (3.2-1) is not subjected to perturbations. In practice, it is not only the initial conditions that are subjected to fluctuations but also the function $\underline{f}$. In modeling various processes, the values of the parameters controlling these processes may not be known exactly and may be subjected to small changes. These will result in small perturbations in $\underline{f}$.

LaSalle and Lefschetz (1961) have discussed **stability under persistent disturbances**. Equation (3.2-1) is modified to

$$\frac{d\underline{x}}{dt} = \underline{f}(\underline{x}, t) + \underline{R}(\underline{x}, t) \qquad (3.2\text{-}22)$$

where $\underline{R}(\underline{x}, t)$ is an unknown function but it is usually assumed that $\| \underline{R}(\underline{x}, t) \|$ is small. The solution $\underline{x}^*$ of Equation (3.2-22) is stable under persistent disturbances if for every $\varepsilon > 0$, there exist two positive numbers δ_1 and δ_2 such that for every $\underline{R}(\underline{x}, t)$ satisfying $\| \underline{R}(\underline{x}, t) \| < \delta_1$ for all $t > t_0$,

$$\| \underline{x}(t_0) - \underline{x}^*(t_0) \| < \varepsilon \quad \Rightarrow \quad \| \underline{x}(t) - \underline{x}^*(t) \| < \delta_2 \text{ for every } t > t_0 \qquad (3.2\text{-}23)$$

Further discussions on this type of stability can be found in Hahn (1963).

In recent years, several authors have discussed the concept of **structural stability** (Arnold, 1983; Rouche et al., 1977; Glendinning, 1994). The system defined by Equation (3.2-1) is structurally stable if the solutions of Equation (3.2-1), with $\underline{f}$ subjected to slight perturbations, are topologically equivalent. Two functions are topologically equivalent if they can be mapped uniquely to each other. In Chapter 1, we have seen that if the equilibrium point of a linear system is an asymptotically stable node, it is also an asymptotically stable node for an almost linear system (Table 1.3-1). This is an example of a structurally stable system. If in the linear system the equilibrium point is a center, no definite conclusion can be drawn for the non-linear system. The equilibrium point can be stable or unstable. That is to say, if the linear system is slightly disturbed, the solutions of the perturbed and unperturbed system can be qualitatively significantly different and the system is structurally unstable.

Other types of stability are considered in Rouche et al. (1977). We shall mainly discuss Liapunov stability and we next describe Liapunov's direct method.

3.3 LIAPUNOV'S DIRECT METHOD – AUTONOMOUS SYSTEM

Many non-linear differential equations cannot be solved exactly and yet it is desirable to establish the stability properties of their solutions. Liapunov's direct method allows us to do this.

Liapunov's direct method is similar to the energy method used in mechanics to investigate the stability of equilibrium points. We recall that we usually consider a system that dissipates energy to be a stable system. In applying Liapunov's direct method, we introduce a function V (**Liapunov function**) which plays a role similar to that of energy. By examining the rate of change of V, we can establish whether the solutions of a differential equation are stable or not.

We first consider an autonomous system. The equations governing the system are written as [Equation (3.2-17b)]

$$\frac{d\underline{x}}{dt} = \underline{f}(\underline{x})$$

(3.3-1a)

$$\underline{f}(\underline{0}) = \underline{0}$$

(3.3-1b)

We now state some definitions.

A function $V(\underline{x})$ is **positive definite** in a region D in the neighborhood of the origin if

(i) $V(\underline{x}) > 0$, for all $\underline{x} \neq \underline{0}$

(3.3-2a)

(ii) $V(\underline{0}) = 0$

(3.3-2b)

A function $V(\underline{x})$ is **positive semi-definite** if

(i) $V(\underline{x}) \geq 0$, for all $\underline{x} \neq \underline{0}$

(3.3-3a)

(ii) $V(\underline{0}) = 0$

(3.3-3b)

A function $V(\underline{x})$ is **negative definite** (or semi-definite) if $-V(\underline{x})$ is positive definite (or semi-definite).

In many applications, we choose V to be a quadratic function and $V(\underline{x})$ can be written as

$$V(\underline{x}) = \underline{x}^{\dagger}\underline{\underline{A}}\,\underline{x}$$

(3.3-4)

where $\underline{\underline{A}}$ is a $(n \times n)$ matrix with elements a_{ij}.

The quadratic expression $V(\underline{x})$ is positive definite (or semi-definite) if the determinants

$$a_{11}, \quad \begin{vmatrix} a_{11} & a_{12} \\ a_{21} & a_{22} \end{vmatrix}, \quad \begin{vmatrix} a_{11} & a_{12} & a_{13} \\ a_{21} & a_{22} & a_{23} \\ a_{31} & a_{32} & a_{33} \end{vmatrix}, \quad ..., \quad |\underline{\underline{A}}|$$

are all positive (or non-negative and $\underline{\underline{A}}$ is singular).

The function $V(\underline{x})$ given by Equation (3.3-4) is negative definite (or semi-definite) if the determinants defined earlier are of alternate signs (or $\underline{A}$ is singular and the values of some of the determinants are zero).

An example of a positive definite function is

$$V(x_1, x_2) = x_1^2 + x_2^2 = [x_1, x_2] \begin{bmatrix} 1 & 0 \\ 0 & 1 \end{bmatrix} \begin{bmatrix} x_1 \\ x_2 \end{bmatrix} \tag{3.3-5a,b}$$

The function

$$V(x_1, x_2) = (x_1^2 - x_2^2)^2 = [x_1, x_2] \begin{bmatrix} 1 & -1 \\ -1 & 1 \end{bmatrix} \begin{bmatrix} x_1 \\ x_2 \end{bmatrix} \tag{3.3-6a,b}$$

is positive semi-definite. It is zero for all $x_1 = x_2$ and x_1, x_2 need not to be zero. Note that, in this case, the determinant is zero.

We assume that $V(\underline{x})$ has continuous first partial derivatives and the derivative of V can be written as

$$\frac{dV}{dt} = \dot{V} = \sum_{i=1}^{n} \frac{\partial V}{\partial x_i} \frac{dx_i}{dt} = \text{grad } V \cdot \dot{\underline{x}} \tag{3.3-7a,b,c}$$

Combining Equations (3.3-1a, 7c) yields

$$\frac{dV}{dt} = \text{grad } V \cdot \underline{f} \tag{3.3-8}$$

Equation (3.3-8) expresses the rate of change of V along a trajectory and can be evaluated. By combining the properties of V and $\dot{V}$, we may specify the stability properties of the system defined by Equations (3.3-1a, b).

Theorem 1

If $V(\underline{x})$ is positive definite and $\dot{V}(\underline{x})$ is negative semi-definite, the equilibrium point $\underline{x} = \underline{0}$ is stable.

Theorem 2

If $V(\underline{x})$ is positive definite and $\dot{V}(\underline{x})$ is negative definite, $\underline{x} = \underline{0}$ is asymptotically stable.

Theorem 3

If $\dot{V}$ is positive definite and $V(\underline{x}) > 0$ at some points in D, $\underline{x} = \underline{0}$ is unstable.

The proofs of Theorems 1 to 3 are given in Hahn (1963). We illustrate the applications of these theorems by considering a few examples.

Example 3.3-1. **Stability of Van der Pol's Equation**

Examine the stability properties of the solution of Van der Pol's equation [Equation (1.4-25)].

For convenience, we reproduce Equation (1.4-25) here

$$\ddot{x} - \varepsilon (1 - x^2) \dot{x} + x = 0 \tag{1.4-25}$$

Equation (1.4-25) can be written as

$$x = x_1, \qquad \dot{x}_1 = x_2 \tag{3.3-9a,b}$$

$$\dot{x}_2 = \varepsilon (1 - x_1^2) x_2 - x_1 \tag{3.3-9c}$$

We note that the linearized system has an isolated critical point at the origin and it is a stable focus.

We now examine the non-linear system. We choose $V(x_1, x_2)$ to be

$$V(x_1, x_2) = x_1^2 + x_2^2 \tag{3.3-10}$$

V is positive definite and its derivative is

$$\dot{V} = 2(x_1 \dot{x}_1 + x_2 \dot{x}_2) \tag{3.3-11}$$

Combining Equations (3.3-9, 11) yields

$$\dot{V} = -2\varepsilon (x_1^2 - 1) x_2^2 \tag{3.3-12}$$

If $x_1^2 < 1$, $\varepsilon < 0$, $\dot{V}$ is negative semi-definite ($\dot{V} = 0$ whenever $x_2 = 0$, $x_1 \neq 0$) and from Theorem 1, we can conclude that the origin is stable.

Example 3.3-2. **Stability of a Conservative System**

Examine the stability of a conservative system via Hamilton's equations of motion.

Hamilton's equations of motion can be written as (Chan Man Fong et al., 1997)

$$\dot{q}^j = \frac{\partial H}{\partial p^j} \tag{3.3-13a}$$

$$\dot{p}^j = -\frac{\partial H}{\partial q^j} \tag{3.3-13b}$$

where p^j and q^j are the generalized momenta and coordinates respectively and H is the Hamiltonian. The Hamiltonian is the total energy of the system and is the sum of the kinetic energy K and the potential energy φ. The kinetic energy K is a quadratic function in $\dot{q}^j$ and is positive definite. The function φ is a function of q^j and not of $\dot{q}^j$ and we choose the level of potential energy such that $\varphi(\underline{0}) = 0$.

A suitable Liapunov function is H (p^j, q^j). We now consider two separate cases.

(i) φ has a minimum at the origin $(q^j = 0, \; \forall \, j)$; it follows that φ is positive in the neighborhood of the origin and H is positive definite. Differentiating yields

$$\dot{H} = \sum_{j=1}^{n} \frac{\partial H}{\partial p^j} \, \dot{p}^j + \sum_{j=1}^{n} \frac{\partial H}{\partial q^j} \, \dot{q}^j \qquad (3.3\text{-}14)$$

Combining Equations (3.3-13a, b, 14), we deduce that $\dot{H} = 0$. From Theorem 1, we conclude that the system is stable, that is, if the potential energy is a minimum at the origin (equilibrium point), the origin is stable.

(ii) φ has a maximum at the origin. We expand φ in a Taylor series about the origin

$$\varphi(q^j) = \varphi(0) + \sum_{j=1}^{n} \frac{\partial \varphi}{\partial q^j} \, q^j + \frac{1}{2} \sum_{i,j=1}^{n} \frac{\partial^2 \varphi}{\partial q^i \, \partial q^j} \, q^i q^j + \dots \qquad (3.3\text{-}15)$$

All the derivatives in Equation (3.3-15) are evaluated at the origin. Since φ has a maximum at the origin, we deduce that

$$\varphi = \frac{\partial \varphi}{\partial q^j} = 0 \quad \text{at } q^j = 0 \qquad (3.3\text{-}16a,b,c)$$

$$\frac{\partial^2 \varphi}{\partial q^i \, \partial q^j} < 0 \quad \text{at } q^j = 0 \qquad (3.3\text{-}16d)$$

In the neighborhood of the origin, φ is a quadratic function and is negative definite. We choose V to be

$$V = \sum_{j=1}^{n} p^j q^j \qquad (3.3\text{-}17)$$

Differentiating and using Equations (3.3-13a, b) yields

$$\dot{V} = \sum_{j=1}^{n} \left(p^j \frac{\partial H}{\partial p^j} - q^j \frac{\partial H}{\partial q^j} \right) \qquad (3.3\text{-}18)$$

The Hamiltonian H is the sum of the kinetic energy $K(p^j)$ and the potential energy $\varphi(q^j)$ and Equation (3.3-18) can be written as

$$\dot{V} = \sum_{j=1}^{n} \left(p^j \frac{\partial K}{\partial p^j} - q^j \frac{\partial \varphi}{\partial q^j} \right) \qquad (3.3\text{-}19)$$

The function K is a quadratic junction in p^j and we have shown that near the origin φ can be approximated as a quadratic function in q^j. On applying Euler's theorem, Equation (3.3-19) becomes

$$\dot{V} = 2K - 2\varphi \qquad (3.3\text{-}20)$$

Since K is positive definite and φ is negative definite, it follows that $\dot{V}$ is positive definite. The function V is positive whenever $q^j > 0$, $p^j > 0$. From Theorem 3, we conclude that the origin is unstable.

We have shown that the equilibrium point is stable (or unstable) if the potential energy at the origin is a minimum (or maximum).

Example 3.3-3. **Asymptotic stability**

Show that the origin of the system

$$\dot{x}_1 = -x_2 - x_1^3 \qquad (3.3\text{-}20a)$$

$$\dot{x}_2 = x_1 - x_2^3 \qquad (3.3\text{-}20b)$$

is asymptotically stable.

We note that for the linearized system, the origin is a center and is stable.

We choose V to be

$$V = x_1^2 + x_2^2 \qquad (3.3\text{-}21)$$

Differentiating and substituting Equations (3.3-20a, b) into the resulting expression yields

$$\dot{V} = -2(x_1^4 + x_2^4) \qquad (3.3\text{-}22)$$

The derivative $\dot{V}$ is negative definite and from Theorem 2, we conclude that the origin is asymptotically stable.

●

In Example 3.3-1, we have only been able to show that the origin is stable whereas the linear system is asymptotically stable. Using the results in Section 1.3, we can conclude that the origin is asymptotically stable. This suggests that we should choose a more appropriate Liapunov function and/or improve Theorems 1–3. Since the publication of Liapunov's work in 1893, numerous improvements have been proposed and they are discussed in Hahn (1963) and Rouche et al. (1977). Here, we state a few theorems without proof.

Theorem 4

If $V(\underline{x})$ is positive definite, $\dot{V}(x)$ is negative definite outside a set of points M, $\dot{V}(x)$ is negative semi-definite on M, and M is not an entire trajectory (except if M is the origin), the origin is asymptotically stable.

Theorem 5

If $V(\underline{x})$ is positive definite, $\dot{V}(x)$ is positive definite outside M (where M is as defined in Theorem 4), and $\dot{V}(x)$ is positive semi-definite on M, $\underline{x} = \underline{0}$ is unstable.

In Example 3.3-1, we noted that $\dot{V}$ is negative definite if $x_1^2 < 1$ and $x_2 \neq 0$ and $\dot{V}$ is zero whenever $x_2 = 0$. The axis $x_2 = 0$ is not a solution of Equations (3.3-9b, c) and is not an entire trajectory. From Theorem 4, we deduce that if $x_1^2 < 1$, the origin is asymptotically stable. This means that if the initial perturbation is such that $\| \underline{x} \| < 1$, the solution will eventually tend to the origin. The region $\| \underline{x} \| < 1$ is the **domain of asymptotic stability**. The domain of asymptotic stability can be identified by determining the regions where V and $\dot{V}$ satisfy the conditions of Theorem 2 or 4. However, in general, this domain depends on the choice of V and for a particular V, the domain needs not be the largest. If the origin is asymptotically stable for all $\underline{x}$, the origin is **globally asymptotically stable** (asymptotically in the large or completely stable).

The following theorems can be used to determine if the origin is globally asymptotically stable.

Theorem 6

If V is positive definite, $V \longrightarrow \infty$ as $\| \underline{x} \| \longrightarrow \infty$ and $\dot{V}$ is negative definite throughout the whole region, $\underline{x} = \underline{0}$ is globally asymptotically stable.

Theorem 7

If V is positive definite, $V \longrightarrow \infty$ as $\| \underline{x} \| \longrightarrow \infty$, $\dot{V}$ is negative definite outside M (M is as defined in Theorem 4), and $\dot{V}$ is negative semi-definite on M, $\underline{x} = \underline{0}$ is globally asymptotically stable.

Example 3.3-4. **Globally Asymptotically Stable**

Show that if the solution $\underline{x} = \underline{0}$ of a linear system is asymptotically stable, it is globally asymptotically stable.

A linear system can be written as [Equation (1.2-4b)]

$$\frac{d\underline{x}}{dt} = \underline{\underline{A}} \, \underline{x} \tag{3.3-23}$$

We choose V to be

$$V = \underline{x}^{\dagger} \, \underline{\underline{B}} \, \underline{x} \tag{3.3-24}$$

where $\underline{\underline{B}}$ is a positive definite symmetric matrix.

Differentiating yields

$$\dot{V} = \dot{\underline{x}}^{\dagger} \, \underline{\underline{B}} \, \underline{x} + \underline{x}^{\dagger} \, \underline{\underline{B}} \, \dot{\underline{x}} \tag{3.3-25}$$

Substituting Equation (3.2-23) into Equation (3.3-25) results in

$$\dot{V} = \underline{x}^{\dagger} \, \underline{\underline{A}}^{\dagger} \, \underline{\underline{B}} \, \underline{x} + \underline{x}^{\dagger} \, \underline{\underline{B}} \, \underline{\underline{A}} \, \underline{x} = \underline{x}^{\dagger} \, \underline{\underline{C}} \, \underline{x} \tag{3.3-26a,b}$$

$$\underline{\underline{C}} = (\underline{\underline{A}}^{\dagger} \, \underline{\underline{B}} + \underline{\underline{B}} \, \underline{\underline{A}}) \tag{3.3-26c}$$

Let $\underline{v}$ and λ be an eigenvector and an eigenvalue of $\underline{\underline{A}}$ respectively. By definition

$$\underline{\underline{A}} \, \underline{v} = \lambda \, \underline{v} \tag{3.3-27}$$

If $\overline{\underline{v}}$ are $\overline{\lambda}$ are the complex conjugates of $\underline{v}$ and λ respectively, the Hermitian of Equation (3.3-27) is ($\underline{\underline{A}}$ is real)

$$\overline{\underline{v}}^{\dagger} \underline{\underline{A}}^{\dagger} = \overline{\lambda} \, \overline{\underline{v}}^{\dagger} \tag{3.3-28}$$

From Equations (3.3-26c, 27, 28), we deduce that

$$\underline{v}^\dagger \underline{C} \; \underline{v} \; = \; \underline{v}^\dagger (\underline{A}^\dagger \underline{B} + \underline{B} \; \underline{A}) \; \underline{v} \; = \; (\underline{v}^\dagger \underline{A}^\dagger \underline{B} \; \underline{v} + \underline{v}^\dagger \underline{B} \; \underline{A} \; \underline{v}) \qquad (3.3\text{-}29a,b)$$

$$= \; (\overline{\lambda} \; \underline{v}^\dagger \underline{B} \; \underline{v} + \underline{v}^\dagger \underline{B} \; \lambda \; \underline{v}) \; = \; (\overline{\lambda} + \lambda) \; \underline{v}^\dagger \underline{B} \; \underline{v} \; = \; 2\lambda_r \; \underline{v}^\dagger \underline{B} \; \underline{v} \qquad (3.3\text{-}29c,d,e)$$

where λ_r is the real part of λ.

The origin is asymptotically stable if the eigenvalues of $\underline{A}$ have negative real parts (Chapter 1) and it follows from Equation (3.3-29e) that, since $\underline{B}$ is positive definite, $\underline{C}$ is negative definite. The function V is a quadratic function of $\underline{x}$ [Equation (3.3-24)] and this implies that $V \longrightarrow \infty$ as $\| \underline{x} \| \longrightarrow \infty$. The derivative $\dot{V}$ is negative definite and the conditions of Theorem 6 are satisfied. We conclude that $\underline{x} = \underline{0}$ is globally asymptotically stable.

Example 3.3-5. **Stability of a chemical reactor**

Examine the stability of a chemical reactor.

We consider the case of a well-stirred reactor in which a homogeneous reaction takes place. The reactants are fed continuously at a rate q with concentration c_0 and temperature T_0. The stability of this system was investigated by Berger and Perlmutter (1964).

A mass and energy balance yields

$$V_0 \frac{dc}{dt} \; = \; - r \, V_0 - q \, (c - c_0) \qquad (3.3\text{-}30a)$$

$$\rho \, V_0 \, C_p \frac{dT}{dt} \; = \; \Delta H \, r \, V_0 - U A \, (T - T_A) - \rho \, q \, C_p \, (T - T_0) \qquad (3.3\text{-}30b)$$

where V_0 is the volume of the reactor, c is the concentration, t is time, r is the reaction rate per unit volume, ρ is the density, C_p is the specific heat, T is the temperature, ΔH is the enthalpy change, U is the overall heat transfer coefficient, and A is the reactor area.

We now introduce the following dimensionless variables

$$y_1 = c/c_0 \, , \quad y_2 = \rho \, C_p T / \Delta H \, c_0 \, , \quad a = \rho \, V_0 C_p \, , \quad b = U A + \rho \, q \, C_p \qquad (3.3\text{-}31a\text{--}d)$$

$$\tau = V_0/q \, , \quad \tau_0 = \rho \, V_0 C_p / U A \qquad (3.3\text{-}31e,f)$$

Substituting Equations (3.3-31 a–f) into Equations (3.3-30a, b) yields

$$\frac{dy_1}{dt} \; = \; -\frac{r}{c_0} - \frac{1}{\tau} \, (y_1 - 1) \qquad (3.3\text{-}32a)$$

$$\frac{dy_2}{dt} = \left(\frac{r}{c_0} - \frac{b}{a}\right) y_2 + \frac{y_{20}}{\tau} + \frac{y_{2A}}{\tau_0} \tag{3.3-32b}$$

The steady state solutions y_{1s} and y_{2s} are given by

$$\frac{dy_{1s}}{dt} = 0 \quad \text{or} \quad y_{1s} = 1 - \tau r_s / c_0 \tag{3.3-33a,b}$$

$$\frac{dy_{2s}}{dt} = 0 \quad \text{or} \quad y_{2s} = (a/b) (r_s/c_0 + y_{20}/\tau + y_{2A}/\tau_0) \tag{3.3-33c,d}$$

where r_s is the steady state reaction rate.

We now write

$$x_1 = y_1 - y_{1s}, \quad x_2 = y_2 - y_{2s}, \quad \sigma = r - r_s \tag{3.3-34a,b,c}$$

Substituting Equations (3.3-34a–c) into Equations (3.3-32a, b) yields

$$\frac{dx_1}{dt} = -\frac{\sigma}{c_0} - \frac{x_1}{\tau} \tag{3.3-35a}$$

$$\frac{dx_2}{dt} = \frac{\sigma}{c_0} - \frac{b}{a} x_2 \tag{3.3-35b}$$

Note that the transformation given by Equations (3.3-34a–c) shifts the steady state solution to the origin.

We now choose V to be

$$V = \left(\frac{\sigma}{c_0} + \frac{x_1}{\tau}\right)^2 + \left(\frac{\sigma}{c_0} - \frac{bx_2}{a}\right)^2 \tag{3.3-36}$$

Differentiating yields

$$\dot{V} = 2\left(\frac{\sigma}{c_0} + \frac{x_1}{\tau}\right)\left(\frac{\dot{\sigma}}{c_0} + \frac{\dot{x}_1}{\tau}\right) + 2\left(\frac{\sigma}{c_0} - \frac{bx_2}{a}\right)\left(\frac{\dot{\sigma}}{c_0} - \frac{b\dot{x}_2}{a}\right) \tag{3.3-37a}$$

$$\dot{\sigma} = \frac{\partial \sigma}{\partial x_1} \dot{x}_1 + \frac{\partial \sigma}{\partial x_2} \dot{x}_2 \tag{3.3-37b}$$

Combining Equations (3.3-35a – 37b) yields

$$\dot{V} = -2 \left[\left(\frac{b}{a} - \frac{1}{c_0} \frac{\partial \sigma}{\partial x_2} \right) \left(\frac{\sigma}{c_0} - \frac{bx_2}{a} \right)^2 + \left(\frac{\sigma}{c_0} - \frac{bx_2}{a} \right) \left(\frac{\sigma}{c_0} + \frac{x_1}{\tau} \right) \left(\frac{1}{c_0} \frac{\partial \sigma}{\partial x_1} - \frac{1}{c_0} \frac{\partial \sigma}{\partial x_2} \right) \right.$$

$$\left. + \left(\frac{1}{\tau} + \frac{1}{c_0} \frac{\partial \sigma}{\partial x_1} \right) \left(\frac{\sigma}{c_0} + \frac{x_1}{\tau} \right)^2 \right] \tag{3.3-38}$$

The square brackets contain quadratic expressions in $(\sigma/c_0 - bx_2/a)$ and $(\sigma/c_0 + x_1/\tau)$. Equation (3.3-38) can be written in the form of Equation (3.3-4) and we identify the elements in the matrix $\underline{\underline{A}}$ to be

$$a_{11} = \left(\frac{b}{a} - \frac{1}{c_0} \frac{\partial \sigma}{\partial x_2} \right), \qquad a_{12} = a_{21} = \frac{1}{2c_0} \left(\frac{\partial \sigma}{\partial x_1} - \frac{\partial \sigma}{\partial x_2} \right) \tag{3.3-39a,b}$$

$$a_{22} = \left(\frac{1}{\tau} + \frac{1}{c_0} \frac{\partial \sigma}{\partial x_1} \right) \tag{3.3-39c}$$

The conditions for $\dot{V}$ to be negative definite are

$$\left(\frac{b}{a} - \frac{1}{c_0} \frac{\partial \sigma}{\partial x_2} \right) > 0 \tag{3.3-40a}$$

$$\left(\frac{b}{a} - \frac{1}{c_0} \frac{\partial \sigma}{\partial x_2} \right) \left(\frac{1}{\tau} + \frac{1}{c_0} \frac{\partial \sigma}{\partial x_1} \right) - \frac{1}{4c_0^2} \left(\frac{\partial \sigma}{\partial x_1} - \frac{\partial \sigma}{\partial x_2} \right)^2 > 0 \tag{3.3-40b}$$

In terms of the original variables, Equations (3.3-40a, b) become

$$b - V_0 \Delta H \left(\partial r / \partial T \right) > 0 \tag{3.3-41a}$$

$$[b - V_0 \Delta H \left(\partial r / \partial T \right)] / a\tau + (b/a)(\partial r / \partial c) - [(V_0 \Delta H / a) \partial r / \partial T + \partial r / \partial c]^2 / 4 > 0 \tag{3.3-41b}$$

The function V as defined by Equation (3.3-36) is positive definite and if the inequalities (3.3-41a, b) are satisfied, the steady state solutions given by Equations (3.3-3b, d) are asymptotically stable. Berger and Perlmutter (1964) have considered the special case where r is given by

$$r = A \exp \left(-Q/T \right) c^n \tag{3.3-42}$$

where A, Q, and n are constants. By assuming numerical values to various quantities, the domain of asymptotic stability can be determined from inequalities (3.3-41a, b).

To apply Liapunov's direct method, we need to find the Liapunov function V and the choice of V depends to an extent on the problem. In many mechanical systems, we can choose V to be the energy of the system. A quadratic function is often a suitable Liapunov function. For some systems, an obvious choice is not available and several methods have been proposed to generate V. We describe three of these methods.

(i) Variable Gradient Method

We recall from Theorem 2 that for the equilibrium point to be asymptotically stable, we require V to be positive definite and $\dot{V}$ to be negative definite. Let $\operatorname{grad} V$ to be of the form

$$
\operatorname{grad} V = \begin{bmatrix} \alpha_{11}x_1 + \alpha_{12}x_2 + \dots + \alpha_{1n}x_n \\ \alpha_{21}x_1 + \alpha_{22}x_2 + \dots + \alpha_{2n}x_n \\ \vdots \\ \alpha_{n1}x_1 + \alpha_{n2}x_2 + \dots + \alpha_{nn}x_n \end{bmatrix} = \begin{bmatrix} g_1 \\ g_2 \\ \vdots \\ g_n \end{bmatrix} \qquad (3.3\text{-}43a,b)
$$

where α_{ij} can be functions of x_i and are chosen such that $\dot{V}$ given by Equation (3.3-8) is negative definite.

The amount of computation can be reduced by noting that $\operatorname{curl} \operatorname{grad} V$ is zero, that is

$$
\frac{\partial g_i}{\partial x_j} - \frac{\partial g_j}{\partial x_i} = 0 \qquad (3.3\text{-}44)
$$

Equation (3.3-43b) yields, on integration

$$
V = \int_0^{x_1} g_1 \, dx_1 + \int_0^{x_2} g_2 \, dx_2 + \dots + \int_0^{x_n} g_n \, dx_n \qquad (3.3\text{-}45)
$$

The next example illustrates this method of constructing V.

Example 3.3-6. **Generating a Liapunov function**

Use the variable gradient method to generate a Liapunov function for the system

$$
\dot{x}_1 = -x_1^5 - 3x_2 \qquad (3.3\text{-}46a)
$$

$$
\dot{x}_2 = x_1^5 - 2x_2 \qquad (3.3\text{-}46b)
$$

Combining Equations (3.3-8, 43a, 46a, b) yields

$$\dot{V} = (\alpha_{11}x_1 + \alpha_{12}x_2)(-x_1^5 - 3x_2) + (\alpha_{21}x_1 + \alpha_{22}x_2)(x_1^5 - 2x_2) \qquad (3.3\text{-}47a)$$

$$= -x_1^6(\alpha_{11} - \alpha_{21}) + x_1x_2[-3\alpha_{11} - 2\alpha_{21} + x_1^4(\alpha_{22} - \alpha_{12})] - x_2^2(3\alpha_{12} + 2\alpha_{22}) \qquad (3.3\text{-}47b)$$

To ensure that $\dot{V}$ is negative definite, we choose α_{ij} such that the coefficient of x_1x_2 is zero and one possible choice is

$$\alpha_{12} = \alpha_{21} = 0, \qquad -3\alpha_{11} + \alpha_{22}x_1^4 = 0 \qquad (3.3\text{-}48a,b,c)$$

Substituting Equations (3.3-48a, b, c) into Equation (3.3-47b) yields

$$\dot{V} = -(1/3)\,\alpha_{22}x_1^{10} - 2\alpha_{22}x_2^2 \qquad (3.3\text{-}49)$$

$\dot{V}$ is negative definite provided α_{22} is positive.

From Equations (3.3-43a, b, 48a–c), we deduce

$$\begin{bmatrix} g_1 \\ g_2 \end{bmatrix} = \begin{bmatrix} (1/3)\,\alpha_{22}x_1^5 \\ \alpha_{22}x_2 \end{bmatrix} \qquad (3.3\text{-}50)$$

Note that Equation (3.3-44) is satisfied, that is

$$\frac{\partial g_1}{\partial x_2} - \frac{\partial g_2}{\partial x_1} = 0 \qquad (3.3\text{-}51)$$

if α_{22} is a constant.

The function V is given by [Equation (3.3-45)]

$$V = \alpha_{22}\left[\frac{1}{3}\int_0^{x_1} x_1^5\,dx_1 + \int_0^{x_2} x_2\,dx_2\right] \qquad (3.3\text{-}52a)$$

$$= \alpha_{22}\,(x_1^6 + 9x_2^2)/18 \qquad (3.3\text{-}52b)$$

The function V is positive definite and $\dot{V}$ is negative definite, and we conclude that the origin is asymptotically stable.

(ii)   Zubov's Method

In Zubov's method, we start by assuming the existence of a function $\phi(\underline{x})$ such that

$$\dot{V} = \text{grad } V \bullet \underline{f} = -\phi(\underline{x}) \qquad (3.3\text{-}53a,b)$$

On integrating Equation (3.3-53b), we obtain V. Equation (3.3-53b) which is a partial differential equation is not easy to solve. We also need to guess $\phi(x)$ and this can be as difficult as guessing V. We use this method to generate a Liapunov function in the next example.

Example 3.3-7. **Construction of a Liapunov function**

Generate a Liapunov function for the system

$$\dot{x}_1 = -2x_1 + 2x_2^4 \qquad (3.3\text{-}54a)$$

$$\dot{x}_2 = -x_2 \qquad (3.3\text{-}54b)$$

We choose ϕ to be

$$\phi = -24(x_1^2 + x_2^2) \qquad (3.3\text{-}55)$$

The function ϕ is negative definite. Equation (3.3-53b) becomes

$$(-2x_1 + 2x_2^4)\frac{\partial V}{\partial x_1} - x_2 \frac{\partial V}{\partial x_2} = -24(x_1^2 + x_2^2) \qquad (3.3\text{-}56)$$

Equation (3.3-56) can be solved by the method of characteristics (Chan Man Fong et al., 1997) and the characteristic equations are

$$\frac{dx_1}{-2x_1 + 2x_2^4} = \frac{dx_2}{-x_2} = \frac{dV}{-24(x_1^2 + x_2^2)} \qquad (3.3\text{-}57a,b)$$

From Equations (3.3-57a), we deduce

$$x_2 \, dx_1 + (2x_1 - 2x_2^4) \, dx_2 = 0 \qquad (3.3\text{-}58)$$

The integrating factor of Equation (3.3-58) is $(1/x_2^3)$ and the solution is

$$x_1 = x_2^2(c - x_2^2) \qquad (3.3\text{-}59)$$

where c is a constant.

Combining Equations (3.3-57b, 59) yields

$$\frac{dV}{dx_2} = 24\,x_2 + 24\,c^2\,x_2^3 - 48\,c\,x_2^5 + 24\,x_2^7 \tag{3.3-60}$$

On integrating, we obtain

$$V = 12\,x_2^2 + 6\,c^2\,x_2^4 - 8\,c\,x_2^6 + 3\,x_2^8 \tag{3.3-61a}$$

$$= 2\,x_1^2 + 12\,x_2^2 + (2\,x_1 + x_2^4)^2 \tag{3.3-61b}$$

The function V is positive definite, $\dot{V}$ is negative definite, and we conclude that the origin is asymptotically stable.

(iii)   Krasovskii's Method

In this method, V is chosen to be

$$V = \underline{f}^\dagger\,\underline{f} \tag{3.3-62}$$

where $\underline{f}$ is the right side of Equation (3.3-1a).

Note that $\underline{f}^\dagger\,\underline{f}$ is $\|\underline{f}\|^2$ and is positive definite. Differentiating yields

$$\dot{V} = \dot{\underline{f}}^\dagger\,\underline{f} + \underline{f}^\dagger\,\dot{\underline{f}} \tag{3.3-63}$$

Using the chain rule, $\dot{\underline{f}}$ can be written as

$$\dot{\underline{f}} = (\operatorname{grad}\underline{f})\,\underline{f} \tag{3.3-64a}$$

Taking the transpose yields

$$\dot{\underline{f}}^\dagger = \underline{f}^\dagger\,(\operatorname{grad}\underline{f})^\dagger \tag{3.3-64b}$$

Combining Equations (3.3-63, 64a, b), we obtain

$$\dot{V} = \underline{f}^\dagger\,[(\operatorname{grad}\underline{f})^\dagger + (\operatorname{grad}\underline{f})]\,\underline{f} \tag{3.3-65a}$$

$$= \underline{f}^\dagger\,\hat{\underline{\underline{J}}}\,\underline{f} \tag{3.3-65b}$$

If $\hat{\underline{\underline{J}}}$ is negative definite, it follows that $\dot{V}$ is negative definite, and the equilibrium point is asymptotically stable. Further, if $V(\underline{x}) \longrightarrow \infty$ as $\| \underline{x} \| \longrightarrow \infty$, the equilibrium point is globally asymptotically stable. Note that $\hat{\underline{\underline{J}}}$ is the sum of the Jacobian of the system $(\mathrm{grad}\,\underline{f})$ and its transpose and is symmetric. In Example 3.3-5, we have used Krasovskii's method to generate V and the elements a_{ij} in Equations (3.3-39 a to c) can be identified to be the elements of $\hat{\underline{\underline{J}}}$. We next solve Example 3.3-7 by Krasovskii's method.

Example 3.3-8. **Comparison of Zubov's and Krasovskii's methods**

Discuss the stability of the system defined by Equations (3.3-54a, b)

In this case, $\underline{f}$ is given by

$$\underline{f} \; = \; \begin{bmatrix} -2x_1 + 2x_2^4 \\[2mm] -x_2 \end{bmatrix} \tag{3.3-66}$$

From Equations (3.3-62, 66), we obtain

$$V \; = \; (-2x_1 + 2x_2^4)^2 + x_2^2 \tag{3.3-67}$$

Note that V is positive definite.

Differentiating $\underline{f}$ yields

$$\hat{\underline{\underline{J}}} \; = \; \begin{bmatrix} -2 & 8x_2^3 \\[2mm] 0 & -1 \end{bmatrix} + \begin{bmatrix} -2 & 0 \\[2mm] 8x_2^3 & -1 \end{bmatrix} = \begin{bmatrix} -4 & 8x_2^3 \\[2mm] 8x_2^3 & -2 \end{bmatrix} \tag{3.3-68a,b}$$

The first minor of $\hat{\underline{\underline{J}}}$ is -4 and is negative. $\hat{\underline{\underline{J}}}$ is negative definite if $\det(\hat{\underline{\underline{J}}})$ is positive, that is

$$8 - 64\,x_2^6 > 0 \tag{3.3-69}$$

Inequality (3.3-69) is satisfied if $8x_2^6 < 1$, that is, only near the origin. The result we obtain here is weaker than that obtained using Zubov's method. In Example 3.3-7, $\dot{V}$ is negative definite for all $\underline{x}$ and the origin is globally asymptotically stable. In the present example, we have been able to deduce only a finite region of asymptotic stability $[\| \underline{x} \| < 1/2]$. Thus, Krasovskii's method is easier to apply but yields weaker results. Improvements of this method have been proposed (Gurel and Lapidus, 1969).

3.4   LIAPUNOV'S DIRECT METHOD – NON-AUTONOMOUS SYSTEM

The theorems stated in the previous section can be extended to non-autonomous systems. In this case, V can be a function of $\underline{x}$ and t. The function $V(\underline{x}, t)$ is **positive definite** if

(i) $V(\underline{x}, t) \geq W(\underline{x}) > 0$ for all $t \geq 0$ and $W(\underline{x})$ is a monotonic increasing function of $\underline{x}$ with $W(\underline{0}) = 0$,

(ii) $V(0, t) = 0$,

(iii) $V(\underline{x}, t)$ is continuous and its first partial derivatives exist.

The function $V(\underline{x}, t)$ is negative definite if $-V$ is positive definite.

$V(\underline{x}, t)$ is **positive semi-definite** if

(i) $V(\underline{x}, t) \geq 0$ for all $t \geq 0$,

(ii) $V(0, t) = 0$,

(iii) $V(\underline{x}, t)$ is continuous and its partial derivatives exist.

An example of a positive definite function is

$$V(x_1, x_2, t) = (x_1^2 + x_2^2)(1 + t^2) \tag{3.4-1}$$

In this case, we can choose W to be

$$W = (x_1^2 + x_2^2) \tag{3.4-2}$$

Conditions (i) to (iii) are satisfied.

The function

$$V(x_1, x_2, t) = (x_1^2 + x_2^2) e^{-t} \tag{3.4-3}$$

is only positive semi-definite because $V \longrightarrow 0$ as $t \longrightarrow \infty$ for finite values of x_1 and x_2.

The derivative $\dot{V}$ is given by

$$\dot{V} = \frac{\partial V}{\partial t} + \sum_{i=1}^{n} \frac{\partial V}{\partial x_i} \frac{dx_i}{dt} \tag{3.4-4a}$$

$$= \frac{\partial V}{\partial t} + \sum_{i=1}^{n} \frac{\partial V}{\partial x_i} f_i \tag{3.4-4b}$$

where f_i are defined by Equation (3.2-1)

Theorems 1, 2, and 3 are valid for both autonomous and non-autonomous systems. Note that in Theorem 2, we require that $\dot{V}$ should be negative definite and this means that $-\dot{V} \geq W(\underline{x}) > 0$ as stated in condition (i). The condition that $\dot{V} < 0$ is not sufficient, as shown in the next example.

Example 3.4-1. Conditions on $V(\underline{x}, t)$

Examine the stability of the following systems via a Liapunov function that depends on t.

(i) $\dot{x}_1 = 0$, $\dot{x}_2 = 0$ \hfill (3.4-5a,b)

(ii) $\dot{x}_1 = -t^2 x_1$, $\dot{x}_2 = -t x_2$ \hfill (3.4-6a,b)

The solutions of Equations (3.4-5a, b) are

$$x_1 = x_{10}, \qquad x_2 = x_{20} \tag{3.4-7a,b}$$

where x_{10} and x_{20} are constants. The solution is stable and not asymptotically stable.

Equations (3.4-6a, b) can be solved and the solution is

$$x_1 = x_{10} \, e^{-t^3/3}, \qquad x_2 = x_{20} \, e^{-t^2/2} \tag{3.4-8a,b}$$

where x_{10} and x_{20} are the initial values of x_1 and x_2. The solution is asymptotically stable.

We choose a Liapunov function defined by

$$V = (x_1^2 + x_2^2)\left[1 + (1+t)^{-1}\right] \tag{3.4-9}$$

Differentiating yields

$$\dot{V} = 2(x_1 \dot{x}_1 + x_2 \dot{x}_2)\left[1 + (1+t)^{-1}\right] - (1+t)^{-2}(x_1^2 + x_2^2) \tag{3.4-10}$$

Substituting Equations (3.4-5a, b) into Equation (3.4-10) yields

$$-\dot{V} = (1+t)^{-2}(x_1^2 + x_2^2) \tag{3.4-11}$$

From Equation (3.4-11), we deduce that $\dot{V}$ is not negative definite because $(1 + t)^{-2} (x_1^2 + x_2^2) \longrightarrow 0$ as $t \longrightarrow 0$ for finite values of t. The condition $-\dot{V} > W(\underline{x})$ is not satisfied.

Combining Equations (3.4-6a, b, 10) yields

$$-\dot{V} = x_1^2 \{2t^2 [1 + (1+t)^{-1}] + (1+t)^{-2}\} + x_2^2 \{2t [1 + (1+t)^{-1}] + (1+t)^{-2}\} > x_1^2 + x_2^2$$

$$(3.4\text{-}12a,b)$$

From Equation (3.4-12a, b), we deduce that $\dot{V}$ is negative definite and it follows that the solution is asymptotically stable.

●

In Example 3.3-5, the criteria for computing the region of asymptotic stability were established without any control. In most cases of practical interest, some mode of control is desirable so as to achieve greater stability as well as better performance. Most mathematical models for various systems, such as mechanical, electrical, economic, and so on, can be described by a system of first order differential equations written as

$$\frac{d\underline{x}}{dt} = \underline{f}(\underline{x}, \underline{u}, t) \tag{3.4-13a}$$

$$\underline{f}(\underline{0}, \underline{0}, t) = 0 \tag{3.4-13b}$$

where $\underline{x}$ arc the **state variables** and $\underline{u}$ are the **control variables**. The number of control variables can be less than or equal to the number of state variables. Equation (3.4-13b) implies that the origin is an equilibrium point.

In an **open-loop system**, the controls are set initially and the system operates without requiring further knowledge of the process developments. This means that the control variables are independent of the state variables.

In a **closed-loop system**, the input of the system is compared with the actual output of the system and the difference between the input and the output is fed back to the system. The object of the control system is to drive the process to the desired output. In this **feedback control system**, the control variables are functions of the state variables.

It is often desired that the system operates in the best possible way and an **index of performance (IP)** is identified. For example, in a chemical reactor, it is desirable to maximize the amount of the final product. An entrepreneur might wish to recoup his investment in the shortest possible time. In this case, the IP is the time required to achieve the goal. In an **optimal control problem**, the controls are adjusted so as to maximize or minimize the index of performance.

The non-linear system defined by Equations (3.4-13a, b) is difficult to analyze. If $\parallel \underline{x} \parallel$ and $\parallel \underline{u} \parallel$ are small, it is possible to linearize the system. Equation (3.4-13a) simplifies to

$$\frac{d\underline{x}}{dt} = \underline{\underline{A}} \, \underline{x} + \underline{\underline{B}} \, u \qquad (3.4\text{-}14)$$

where $\underline{\underline{A}}$ is a $(n \times n)$ matrix, $\underline{\underline{B}}$ is a $(m \times n)$ matrix with $(n \geq m)$, and $\underline{\underline{A}}$ and $\underline{\underline{B}}$ can be functions of time.

For a closed-loop system, $\underline{u}$ is a linear function of $\underline{x}$ and Equation (3.4-14) becomes

$$\frac{d\underline{x}}{dt} = \underline{\underline{A}} \, \underline{x} + \underline{\underline{K}} \, \underline{x} \qquad (3.4\text{-}15)$$

where $\underline{\underline{K}}$ is the control matrix.

In the next example, we consider an open-loop linear system.

Example 3.4-2. **RLC circuit**

Examine the stability of a circuit consisting of a voltage source, a capacitor, an inductor, and a resistor connected in series.

The equation for the system described here is

$$L \frac{d^2 i}{dt^2} + R \frac{di}{dt} + \frac{i}{C} = \frac{de_i}{dt} \qquad (3.4\text{-}16)$$

where L is the inductance, R is the resistance, C is the capacitance, i is the current, and e_i is the voltage source.

A detailed derivation of Equation (3.4-16) is given in Close and Frederick (1978).

Equation (3.4-16) is now written as a system of first order equations. We write

$$x_1 = i \qquad (3.4\text{-}17a)$$

$$\frac{dx_1}{dt} = x_2 \qquad (3.4\text{-}17b)$$

$$\frac{dx_2}{dt} = -\frac{R}{L} x_2 - \frac{x_1}{LC} + \frac{\dot{e}_i}{L} \qquad (3.4\text{-}17c)$$

The state variables are x_1 and x_2 and the control variable is $(\dot{e}_i / L)$. Note that x_1 is the current (i) and Lx_2 is the voltage across the inductor. We assume that R, L, and C are constants. Comparing Equations (3.4-14, 17 a-c), we identify

$$\underline{A} = \begin{bmatrix} 0 & 1 \\ -1/LC & -R/L \end{bmatrix}, \qquad \underline{B} = \begin{bmatrix} 0 & 0 \\ 0 & 1 \end{bmatrix}, \qquad \underline{u} = \begin{bmatrix} 0 \\ \dot{e}_i/L \end{bmatrix} \qquad \text{(3.4-18a–c)}$$

We first investigate the stability of the system without control $(\dot{e}_i/L = u = 0)$. Since the system is linear, we choose a Liapunov function which is quadratic in $\underline{x}$, that is

$$V = \underline{x}^\dagger \underline{P} \, \underline{x} \qquad \text{(3.4-19a)}$$

$$\underline{P} = \begin{bmatrix} R/L+2/RC & 1 \\ \\ 1 & 2L/R \end{bmatrix} \qquad \text{(3.4-19b)}$$

The derivative $\dot{V}$ is given by

$$\dot{V} = \dot{\underline{x}}^\dagger \underline{P} \, \underline{x} + \underline{x}^\dagger \underline{P} \, \dot{\underline{x}} \qquad \text{(3.4-20)}$$

Combining Equations (3.4-18a, 20) yields

$$\dot{V} = \underline{x}^\dagger \underline{A}^\dagger \underline{P} \, \underline{x} + \underline{x}^\dagger \underline{P} \, \underline{A} \, \underline{x} \qquad \text{(3.4-21a)}$$

$$= \underline{x}^\dagger \underline{Q} \, \underline{x} \qquad \text{(3.4-21b)}$$

$$\underline{Q} = \begin{bmatrix} -2/LC & 0 \\ \\ 0 & -2 \end{bmatrix} \qquad \text{(3.4-21c)}$$

The matrix $\underline{P}$ is positive definite and $\underline{Q}$ is negative definite and we conclude that the system is asymptotically stable.

We now include the control. Combining Equations (3.4-18a-c, 20) yields

$$\dot{V} = \underline{x}^\dagger (\underline{A}^\dagger \underline{P} + \underline{P} \, \underline{A}) \, \underline{x} + \underline{u}^\dagger \underline{B}^\dagger \underline{P} \, \underline{x} + \underline{x}^\dagger \underline{P} \, \underline{B} \, \underline{u} \qquad \text{(3.4-22a)}$$

$$= -2x_1^2/LC - 2x_2^2 + x_1 u + 2Lx_2 u/R \qquad \text{(3.4-22b)}$$

$$= -\left\{(2/LC)\,[(x_1 - uLC/4)^2] + 2\,[(x_2 - Lu/2R)^2] - LCu^2/8 - L^2u^2/2R^2\right\} \qquad \text{(3.4-22c)}$$

In this case, $\dot{V}$ is not negative definite and $\dot{V}$ is positive when

$$x_1 = uLC/4, \qquad x_2 = uL/2R \qquad\qquad (3.4\text{-}23a,b)$$

The system is unstable if $u \neq 0$. This is physically plausible because the source of voltage will drive the current away from the equilibrium.

●

In deriving Equation (3.4-15), we have assumed that it is possible to measure all the n state variables $(x_1, ..., x_n)$. In reality, it is often impossible to measure all state variables $(x_1, ..., x_n)$. Instead, a feedback control involving r output variables $(y_1, ..., y_r)$ with $r < n$ are measured. For a linear output feedback system, we write

$$\underline{y} = \underline{\underline{C}}\ \underline{x} \qquad\qquad (3.4\text{-}24a)$$

$$\underline{u} = \underline{\underline{K}}\ \underline{y} \qquad\qquad (3.4\text{-}24b)$$

where $\underline{\underline{C}}$ is a $(r \times n)$ matrix and $\underline{\underline{K}}$ is a $(m \times r)$ matrix.

In a non-linear system, the matrices $\underline{\underline{C}}$ and $\underline{\underline{K}}$ in Equations (3.4-24a, b) are replaced by non-linear functions. In a **direct control system**, the control variables u_i are expressed as functions of the output variables y_j. In an **indirect control system**, the $\dot{u}_i$ are expressed as functions of y_i. The next example considers an indirect control system.

Example 3.4-3. **Bulgakov problem**

Examine the stability of an oscillatory system subject to an indirect control.

This problem is discussed in detail in Letov (1961). The system is an oscillatory system with a phase angle ϕ and a control variable u. The equation governing the system can be written as

$$T^2\ddot{\phi} + U\dot{\phi} + u = 0 \qquad\qquad (3.4\text{-}25)$$

where T^2 is the inertia and U is the damping in the regulated system. The desired position of the system is $\phi = 0$ and the object of the control u is to ensure that the system returns to $\phi = 0$ if ever it departs from the equilibrium position.

We consider the case of an indirect control and u is given by

$$\dot{u} = f^*(y) \qquad\qquad (3.4\text{-}26a)$$

$$y = a\,\phi + b\,\dot{\phi} + c\,\ddot{\phi} - du \qquad\qquad (3.4\text{-}26b)$$

where a to d are the constants of the regulator and f^* is a non-linear function of y. It is assumed that f^* satisfies the following conditions

(i) $\displaystyle\int_0^s f^*(y)\,dy \longrightarrow \infty$ as $|s| \longrightarrow \infty$

(3.4-27a)

(ii) $f^*(y) = 0$, $|y| \le y_0$

(3.4-27b)

Equation (3.4-27b) implies that the control is not operational whenever the magnitude of the output variable is less than y_0.

We introduce the following new variables

$$k = T^2/(c+T^2 d),\quad r = k/T^2,\quad \tau = \sqrt{r}\,t,\quad p_1 = a,$$

$$p_2 = [b-cU^2/T^2]\sqrt{r},\quad \phi = \eta_1,\quad d\phi/dt = \sqrt{r}\,\eta_2,\quad a_{22} = -U/(\sqrt{r}\,T^2)$$

$$\zeta = u/k,\quad f(y) = f^*(y)/(\sqrt{r}\,k)$$

(3.4-28)

Using Equations (3.4-28), Equations (3.4-25, 26a, b) become

$$\frac{d\eta_1}{d\tau} = \eta_2$$

(3.4-29a)

$$\frac{d\eta_2}{d\tau} = a_{22}\,\eta_2 - \zeta$$

(3.4-29b)

$$\frac{d\zeta}{d\tau} = f(y)$$

(3.4-29c)

$$y = p_1\eta_1 + p_2\eta_2 - \zeta$$

(3.4-29d)

The steady state solution is given by

$$\eta_2 = 0,\qquad \zeta = 0$$

(3.4-30a,b)

From Equation (3.4-27b), we deduce that the control does not operate if

$$|\eta_1| < y_0/p_1$$

(3.4-31)

For convenience, we introduce x_1 and x_2 and they are defined by

$$x_1 = \zeta,\qquad x_2 = \zeta - a_{22}\eta_2$$

(3.4-32a,b)

We differentiate Equation (3.4-29d) with respect to τ and combine the resulting expression with Equations (3.4-29a-d, 32a, b), we obtain

$$\frac{dx_1}{d\tau} = f(y) \qquad \text{(3.4-33a)}$$

$$\frac{dx_2}{d\tau} = -\alpha x_2 + f(y) \qquad \text{(3.4-33b)}$$

$$\frac{dy}{d\tau} = \beta_1 x_1 + \beta_2 x_2 - f(y) \qquad \text{(3.4-33c)}$$

where

$$\alpha = -a_{22} = U/(\sqrt{r}\,T^2), \quad \beta_1 = -p_1/\alpha, \quad \beta_2 = (p_1 - \alpha p_2)/\alpha \qquad \text{(3.4-33d-g)}$$

To examine the stability of Equations (3.4-33a-c), we choose a Liapunov function V of the form

$$V = \frac{1}{2}[\alpha_1 x_1^2 + (\alpha_2^2/\alpha) x_2^2] + \int_0^y f(y)\,dy \qquad \text{(3.4-34)}$$

Differentiating yields

$$\frac{dV}{d\tau} = \alpha_1 x_1 \frac{dx_1}{d\tau} + (\alpha_2^2/\alpha) x_2 \frac{dx_2}{d\tau} + f(y)\frac{dy}{d\tau} \qquad \text{(3.4-35a)}$$

$$= -[\alpha_2^2 x_2^2 + f^2(y)] + f(y)[x_1(\alpha_1 + \beta_1) + x_2(\alpha_2^2/\alpha + \beta_2)] \qquad \text{(3.4-35b)}$$

We need to establish the sign of V and $dV/d\tau$. To simplify the expression in Equation (3.4-35b), we choose

$$\alpha_1 + \beta_1 = 0, \qquad \alpha_2^2/\alpha + \beta_2 + 2\alpha_2 = 0 \qquad \text{(3.4-36a,b)}$$

Combining Equations (3.4-35b, 36b) yields

$$\frac{dV}{d\tau} = -(\alpha_2^2 x_2^2 + 2\alpha_2 x_2 f + f^2) \qquad \text{(3.4-37a)}$$

$$= -(\alpha_2 x_2 + f)^2 \qquad \text{(3.4-37b)}$$

Equation (3.4-37b) shows that $dV/d\tau$ is positive semi-definite if α_2 is real. From Equation (3.4-36b), we deduce that α_2 is real if

$$\alpha^2 - \beta_2 \alpha > 0 \quad \text{or} \quad \alpha(\alpha - \beta_2) > 0 \quad \text{or} \quad \alpha > 0 \text{ and } \alpha > \beta_2 \qquad \text{(3.4-38a-d)}$$

The function V is positive definite if $\alpha_1 > 0$ and $\alpha > 0$. From Equations (3.4-33f, 36a), it is seen that $\alpha_1 > 0$ if $\alpha > 0$ and $a\,(= p_1) > 0$. The system is stable if

$$\alpha > 0, \quad a > 0, \quad \alpha > \beta_2 \qquad\qquad (3.4\text{-}39a\text{--}c)$$

In terms of the original variables, inequalities (3.4-39a-c) can be expressed as

$$U > 0, \quad a > 0, \quad U\,(Ud + b)/T^2 > a \qquad\qquad (3.4\text{-}40a\text{--}c)$$

The condition $U > 0$ implies that the system is damped and not excited.

●

3.5 HYDRODYNAMIC STABILITY

It was observed by Reynolds (1883) that the laminar flow in a circular smooth pipe breaks down when the Reynolds number (Re) exceeds a critical value Re_c. This means that when $Re > Re_c$, the flow becomes unstable. Since then, considerable progress has been made in hydrodynamic stability and some of the recent achievements are reviewed in Swinney and Gollub (1981) and Hsieh and Ho (1994).

In this section, we consider the stability of the flow between two coaxial cylinders. This problem was first considered by Taylor (1923). He observed and calculated the critical rotational speed at which the flow became unstable. His calculations and observations were in complete agreement and marked the first success in the theory of hydrodynamic stability.

We choose the usual cylindrical polar coordinates (r, θ, z) with z-axis coinciding with the axis of the cylinders. We denote the radii of the outer and inner cylinders by R_0 and R_i respectively. We assume the outer cylinder to be at rest and the inner cylinder to rotate at a speed Ω.

To examine the stability of the flow, we first determine the basic flow. The velocity distribution is

$$v_r = 0, \quad v_\theta = V(r), \quad v_z = 0 \qquad\qquad (3.5\text{-}1a\text{--}c)$$

where v_r, v_θ, and v_z are the velocity components in the r, θ, and z directions respectively.

To obtain the explicit form of $V(r)$, we have to solve the Navier-Stokes equations which can be written as

$$\rho\left(\frac{Dv_r}{Dt} - \frac{v_\theta^2}{r}\right) = -\frac{\partial p}{\partial r} + \eta\left(\nabla^2 v_r - \frac{v_r}{r^2} - \frac{2}{r^2}\frac{\partial v_\theta}{\partial \theta}\right) \qquad\qquad (3.5\text{-}2a)$$

$$\rho \left(\frac{Dv_\theta}{Dt} + \frac{v_r v_\theta}{r} \right) = -\frac{1}{r} \frac{\partial p}{\partial \theta} + \eta \left(\nabla^2 v_\theta - \frac{v_\theta}{r^2} + \frac{2}{r^2} \frac{\partial v_r}{\partial \theta} \right) \qquad (3.5\text{-}2b)$$

$$\rho \left(\frac{Dv_z}{Dt} \right) = -\frac{\partial p}{\partial z} + \eta \nabla^2 v_z \qquad (3.5\text{-}2c)$$

$$\frac{D}{Dt} = \frac{\partial}{\partial t} + v_r \frac{\partial}{\partial r} + \frac{v_\theta}{r} \frac{\partial}{\partial \theta} + v_z \frac{\partial}{\partial z} \qquad (3.5\text{-}2d)$$

$$\nabla^2 = \frac{\partial^2}{\partial r^2} + \frac{1}{r} \frac{\partial}{\partial r} + \frac{1}{r^2} \frac{\partial^2}{\partial \theta^2} + \frac{\partial^2}{\partial z^2} \qquad (3.5\text{-}2e)$$

where ρ is the density, p is the pressure, and η is the viscosity.

The no-slip boundary conditions imply

$$V(R_0) = 0, \qquad V(R_i) = R_i \Omega \qquad (3.5\text{-}3a,b)$$

We assume the flow to be axi-symmetric and from Equations (3.5-1a-c, 2c), we deduce that p is a function of r only and we write

$$p = P(r) \qquad (3.5\text{-}4)$$

Combining Equations (3.5-1a-c, 2a, b, 4) yields

$$-\rho \frac{V^2}{r} = -\frac{dP}{dr} \qquad (3.5\text{-}5a)$$

$$\frac{d^2V}{dr^2} + \frac{1}{r} \frac{dV}{dr} - \frac{V}{r^2} = 0 \qquad (3.5\text{-}5b)$$

The solution of Equation (3.5-5b) subject to Equations (3.5-3a, b) is

$$V = Ar + B/r \qquad (3.5\text{-}6a)$$

$$A = \Omega R_i / (R_i^2 - R_0^2) \qquad (3.5\text{-}6b)$$

$$B = \Omega R_i R_0^2 / (R_0^2 - R_i^2) \qquad (3.5\text{-}6c)$$

The pressure P can be obtained by combining Equations (3.5-5a, 6a).

To examine the stability of the flow, we superimpose on this flow a disturbance. If the disturbance grows with time, the flow is unstable and if the disturbance decays to zero, the flow is stable. The velocity and the pressure distributions of the perturbed flow are written as

$$v_r = u, \quad v_\theta = V(r) + v, \quad v_z = w, \quad p = P(r) + p' \tag{3.5-7a-d}$$

We consider the case of linear stability and axi-symmetric disturbances. The quantities u, v, w, and p' are expressed as

$$u = u_1(r, z) e^{\sigma t}, \quad v = v_1(r, z) e^{\sigma t}, \quad w = w_1(r, z) e^{\sigma t}, \quad p' = p_1(r, z) e^{\sigma t} \tag{3.5-8a-d}$$

where $\sigma \; (= \sigma_r + i\sigma_i)$ is a complex quantity.

If σ_r is positive, the flow is unstable and if σ_r is negative, the flow is stable.

Substituting Equations (3.5-8a-d) into Equations (3.5-2a-c) and linearizing yields

$$\rho \left[\sigma u_1 - 2v_1 (A + B/r^2) \right] = -\frac{\partial p_1}{\partial r} + \eta \left(\nabla^2 u_1 - u_1/r^2 \right) \tag{3.5-9a}$$

$$\rho \left(\sigma v_1 + 2u_1 A \right) = \eta \left(\nabla^2 v_1 - v_1/r^2 \right) \tag{3.5-9b}$$

$$\rho \sigma w_1 = -\frac{\partial p_1}{\partial z} + \eta \nabla^2 w_1 \tag{3.5-9c}$$

The equation of continuity is

$$\frac{\partial u_1}{\partial r} + \frac{u_1}{r} + \frac{\partial w_1}{\partial z} = 0 \tag{3.5-10}$$

In many experimental situations, the gap between the cylinders is small and Equations (3.5-9a-c, 10) can be greatly simplified if the small gap approximation is made.

If we denote the gap between the cylinders $(R_0 - R_i)$ by d, the small gap approximation implies that d/R_i is small. We make a change of variables and write

$$r = R_i + xd, \quad z = z_1 d \tag{3.5-11a,b}$$

Using the chain rule, we obtain

$$\frac{\partial u_1}{\partial r} = \frac{\partial u_1}{\partial x} \frac{dx}{dr} = \frac{1}{d} \frac{\partial u_1}{\partial x} \tag{3.5-12a,b}$$

$$\frac{\partial^2 u_1}{\partial r^2} = \frac{1}{d} \frac{\partial}{\partial x} \left(\frac{1}{d} \frac{\partial u_1}{\partial x} \right) = \frac{1}{d^2} \frac{\partial^2 u_1}{\partial x^2} \tag{3.5-12c,d}$$

The quantities $1/r$ and $1/r^2$ can be approximated as

$$\frac{1}{r} = \frac{1}{R_i} \left(1 + \frac{xd}{R_i}\right)^{-1} = \frac{1}{R_i} \left(1 - \frac{xd}{R_i} + ...\right) \tag{3.5-13a,b}$$

$$\frac{1}{r^2} = \frac{1}{R_i^2} \left(1 + \frac{xd}{R_i}\right)^{-2} = \frac{1}{R_i^2} \left(1 - \frac{2xd}{R_i} + ...\right) \tag{3.5-13c,d}$$

Similar expressions can be obtained for the other quantities.

Equations (3.5-9a-c, 10) simplify to

$$\rho \left[\sigma u_1 - 2\Omega v_1 (1-x)\right] = -\frac{1}{d} \frac{\partial p_1}{\partial x} + \frac{\eta}{d^2} \left(\frac{\partial^2 u_1}{\partial x^2} + \frac{\partial^2 u_1}{\partial z_1^2}\right) \tag{3.5-14a}$$

$$\rho \left(\sigma v_1 - \frac{R_i \Omega u_1}{d}\right) = \frac{\eta}{d^2} \left(\frac{\partial^2 v_1}{\partial x^2} + \frac{\partial^2 v_1}{\partial z_1^2}\right) \tag{3.5-14b}$$

$$\rho \sigma w_1 = -\frac{1}{d} \frac{\partial p_1}{\partial z_1} + \frac{\eta}{d^2} \left(\frac{\partial^2 w_1}{\partial x^2} + \frac{\partial^2 w_1}{\partial z_1^2}\right) \tag{3.5-14c}$$

$$\frac{\partial u_1}{\partial x} + \frac{\partial w_1}{\partial z_1} = 0 \tag{3.5-14d}$$

From Equation (3.5-14d), we deduce that a stream function $\psi(x, z_1)$ can be introduced and u_1, w_1 are given by

$$u_1 = \partial \psi / \partial z_1, \qquad w_1 = -\partial \psi / \partial x \tag{3.5-15a,b}$$

Eliminating p_1 between Equations (3.5-14a, c) and substituting Equations (3.5-12a, b) into the resulting expression yields

$$\rho \left[\sigma \nabla^2 \psi - 2\Omega (1-x) (\partial v_1 / \partial z_1)\right] = (\eta/d^2) \nabla^4 \psi \tag{3.5-16a}$$

$$\nabla^2 \psi = \partial^2 \psi / \partial x^2 + \partial^2 \psi / \partial z_1^2 \tag{3.5-16b}$$

We assume that the cylinders are of infinite length and that (ψ, v_1) are periodic in the axial direction. To non-dimensionalize Equations (3.5-14b, 16a), we write ψ and v_1 as

$$\psi = -\text{Re } R_i \Omega \chi(x) \cos \lambda z_1, \qquad v = [R_i \Omega \varphi(x) \sin \lambda z_1]/2\lambda \tag{3.5-17a,b}$$

where Re $(= d^2 \rho \Omega_1 / \eta)$ is the Reynolds number and λ is the non-dimensional wave number of the disturbance.

Substituting Equations (3.5-17a, b) into Equations (3.5-14b, 16a) yields respectively

$$c\phi - T\lambda^2\chi = (D^2 - \lambda^2)\,\phi \qquad\qquad (3.5\text{-}18a)$$

$$c(D^2 - \lambda^2)\chi + (1 - x)\,\phi = (D^2 - \lambda^2)^2\chi \qquad\qquad (3.5\text{-}18b)$$

where $c\ (= \sigma Re/\Omega)$ is a dimensionless number, $T\ (= 2\,Re^2 R_i/d)$ is the Taylor number, and $D \equiv d/dx$.

The no-slip boundary conditions are given by

$$\chi = D\chi = \phi = 0 \quad \text{at } x = 0 \text{ and } x = 1 \qquad\qquad (3.5\text{-}19a\text{-}c)$$

Equations (3.5-18a – 19c) define an eigenvalue problem. It is stated earlier that the stability of the flow depends on the sign of the real part of $c\ (c_r)$. The values of T that correspond to $c_r = 0$ demarcate the region of stability from that of instability and they depend on λ.

The curve of these values of T versus λ is the **neutral stability curve** and a typical one is shown in Figure 3.5-1. If the imaginary part of $c\ (c_i)$ is non-zero, then the disturbances are oscillatory in time. It has been observed experimentally that when the laminar flow breaks down, the disturbances are steady and these steady cells are **Taylor vortices**. Thus, both the real and imaginary parts of c are zero and this is the **principle of exchange of stabilities** (Chandrasekhar, 1961).

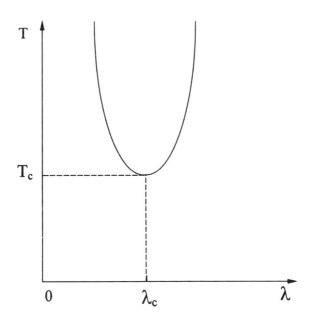

FIGURE 3.5-1   Neutral stability curve

Setting $c = 0$, Equations (3.5-18a, b) simplify to

$$-T\lambda^2 \chi = (D^2 - \lambda^2)\,\varphi \tag{3.5-20a}$$

$$(1 - x)\,\varphi = (D^2 - \lambda^2)^2\,\chi \tag{3.5-20b}$$

Chandrasekhar (1961) solved the eigenvalue problem by the Galerkin method and we adopt his method. The function φ is expanded in a Fourier series that satisfies the boundary conditions, that is, φ is expressed as

$$\varphi = \sum_{m=1}^{\infty} b_m \sin m\pi x \tag{3.5-21}$$

Substituting Equation (3.5-21) into Equation (2.5-20b) yields

$$(D^2 - \lambda^2)^2\,\chi = (1 - x) \sum_{m=1}^{\infty} b_m \sin m\pi x \tag{3.5-22}$$

The solution of this non-homogeneous equation that satisfies the boundary conditions [Equations (3.5-19a, b)] is

$$\chi = \sum_{m=1}^{\infty} \chi_m \tag{3.5-23a}$$

$$\chi_m = [b_m/(m^2\pi^2 + \lambda^2)]\,\{A_1^m \cosh \lambda x + B_1^m \sinh \lambda x + A_2^m\, x \cosh \lambda x + B_2^m\, x \sinh \lambda x$$

$$+ (1 - x)\sin m\pi x - [4m\pi/(m^2\pi^2 + \lambda^2)]\cos m\pi x\} \tag{3.5-23b}$$

$$A_1^m = 4m\pi / (m^2\pi^2 + \lambda^2) \tag{3.5-23c}$$

$$A_2^m = -[m\pi / (\sinh^2 \lambda - \lambda^2)]\,[\sinh^2 \lambda + \beta_m \lambda\,(\sinh \lambda + \lambda \cosh \lambda) - \gamma_m \lambda \sinh \lambda] \tag{3.5-23d}$$

$$B_1^m = [m\pi / (\sinh^2 \lambda - \lambda^2)]\,[\lambda + \beta_m(\sinh \lambda + \lambda \cosh \lambda) - \gamma_m \sinh \lambda] \tag{3.5-23e}$$

$$B_2^m = [m\pi / (\sinh^2 \lambda - \lambda^2)]\,[\sinh \lambda \cosh \lambda - \lambda + \beta_m \lambda^2 \sinh \lambda - \gamma_m(\lambda \cosh \lambda - \lambda \sinh \lambda)] \tag{3.5-23f}$$

$$\beta_m = [-4 / (m^2\pi^2 + \lambda^2)]\,[\cosh \lambda + (-1)^{m+1}] \tag{3.5-23g}$$

$$\gamma_m = -4\lambda \sinh\lambda / (m^2\pi^2 + \lambda^2) \tag{3.5-23h}$$

Substituting Equations (3.5-21, 23a) into Equation (3.5-20a) yields

$$\sum_{m=1}^{\infty} [b_m (m^2\pi^2 + \lambda^2) \sin m\pi x + T\lambda^2 \chi_m] = 0 \tag{3.5-24}$$

Multiplying Equation (3.5-24) by $\sin n\pi x$ and integrating the resulting expressions with respect to x yields a system of linear algebraic equations of the form

$$\sum_{m=1}^{\infty} [b_m (m^2\pi^2 + \lambda^2)] D_{mn} = 0, \qquad n = 1, 2, ... \tag{3.5-25a}$$

$$D_{mn} = 4mn\pi^2 [1 - (-1)^{m+n}] / (n^2\pi^2 + \lambda^2)(m^2\pi^2 + \lambda^2) + (1/2)\delta_{nm} - X_{mn}$$

$$- [2\lambda mn\pi^2 / (n^2\pi^2 + \lambda^2)(\sinh^2\lambda - \lambda^2)] \{(\sinh\lambda \cosh\lambda - \lambda)$$

$$+ (-1)^{n+1}(\sinh\lambda - \lambda\cosh\lambda) + 4\lambda\sinh\lambda [\sinh\lambda + (-1)^{m+1}\lambda]$$

$$[(-1)^{n+m} - 1] / (m^2\pi^2 + \lambda^2)\} - (1/2)(n^2\pi^2 + \lambda^2)^3 \delta_{nm} / \lambda^2 T \tag{5.3-25b}$$

$$X_{mn} = \begin{cases} 0, & m+n \text{ is even}, m \neq n \\ 1/4, & m = n \\ 4mn[2/(m^2+n^2) - 1/\pi^2 (n^2 - m^2)]/(n^2 - m^2), & m+n \text{ is odd} \end{cases} \tag{3.5-25c}$$

δ_{nm} is the Kronecker delta.

Writing out Equation (3.5-25a) for two cases $(m = 1, n = 1)$ and $(m = 2, n = 2)$ yields

(a) $m = n = 1$

$$b_1 D_{11} / (\pi^2 + \lambda^2) = 0 \tag{3.5-26}$$

Since b_1 is non-zero, it follows that D_{11} is zero and from Equation (3.5-25b), we can calculate T as a function of λ. The critical Taylor number T_c corresponds to the minimum value of T and we denote the corresponding value of λ by λ_c (see Figure 3.5-1).

(b)   m = n = 2

$$b_1 D_{11} / (\pi^2 + \lambda^2) + b_2 D_{21} / (4\pi^2 + \lambda^2) = 0 \qquad \text{(3.5-27a)}$$

$$b_1 D_{12} / (\pi^2 + \lambda^2) + b_2 D_{22} / (4\pi^2 + \lambda^2) = 0 \qquad \text{(3.5-27b)}$$

The coefficients b_1 and b_2 are non-zero and for a non-trivial solution, we require the determinant

$$\begin{vmatrix} D_{11}/(\pi^2 + \lambda^2) & D_{21}/(4\pi^2 + \lambda^2) \\ D_{12}/(\pi^2 + \lambda^2) & D_{22}/(4\pi^2 + \lambda^2) \end{vmatrix} = 0 \qquad \text{(3.5-28)}$$

Equation (3.5-28) allows for the calculation of T_c and λ_c.

In case (a), $T_c = 3430$, $\lambda_c = 3.12$, and in case (b), $T_c = 3404$, $\lambda_c = 3.12$. It can be seen that this method of solving eigenvalue problems (Galerkin method) is quite powerful and the first approximation (m = n = 1) gives an acceptable value of T_c. Other methods have been used to solve the eigenvalue problem (Drazin and Reid, 1981) and the results are in agreement.

The present analysis has shown that if $T < T_c$, the flow is purely azimuthal and is given by Equations (3.5-1a-c, 6a-c). At $T = T_c$, there is another possible flow given by Equations (3.5-7a-c) with non-vanishing (u, v, w). This secondary flow is steady (principle of exchange of stabilities) and appears as Taylor vortices. This means that there is a bifurcation point at $T = T_c$. As T is further increased, there is another bifurcation point at which the Taylor vortices become unstable and a new flow pattern emerges. We now have a wavy vortex flow. On further increasing T, the flow becomes more random and eventually the flow is turbulent. Coles (1965) and Andereck et al. (1986) have reported experimental observations for a wide range of Reynolds number for both co-rotating and counter-rotating cylinders. Note that in the case of co-rotating cylinders, the basic azimuthal velocity $V(r)$ is of one sign throughout the gap between the cylinders whereas in the case of counter-rotating cylinders, $V(r)$ changes sign and is zero for some values of r. This difference gives rise to different forms of instability.

Non-linear stability analyses have successfully described the growth of Taylor vortices and the onset of wavy vortex flows. Beyond this point, the Navier equations are too complicated. Instead model equations are used to explain the experimental observations and several models have been proposed to describe the transition from laminar flow to chaos (Swinney and Gollub, 1981, Example 4.2-1). Chaos will be considered in Chapter 5.

PROBLEMS

1.   Verify that

$$x_1 = a \cos t, \qquad x_2 = a \sin t$$

and

$$x_1^* = a \cos (t + \delta), \qquad x_2^* = a \sin (t + \delta)$$

are solutions of the system

$$\dot{x}_1 = -x_2 + x_1 (x_1^2 + x_2^2 - a^2)$$

$$\dot{x}_2 = x_1 + x_2 (x_1^2 + x_2^2 - a^2)$$

Calculate the Euclidean norms

$$\| \underline{x}(0) - \underline{x}^*(0) \| \quad \text{and} \quad \| \underline{x}(t) - \underline{x}^*(t) \|$$

and deduce the stability of the solutions.

Show that the differential equations of the system, in polar coordinates, can be written as

$$\dot{r} = r (r^2 - a)$$

$$\dot{\theta} = 1$$

Obtain the solution and sketch the trajectories. Is the periodic solution $r = a$ orbitally stable?

2.   Solve the following equations

(a) $\dot{x} = 0, \qquad x(0) = 1$

(b) $\dot{x} = -x, \qquad x(0) = 1$

Examine the stability of the solutions.

Disturb the system slightly. The equations governing the perturbed systems (a) and (b) are written as

(a') $\dot{x} = \varepsilon x^2, \qquad x(0) = 1$

(b') $\dot{x} = -x + \varepsilon x^2$, $x(0) = 1$

where $\varepsilon \ll 1$.

Obtain the solutions for both cases. Are the systems structurally stable?

3.   (a)   Show that the system

$$\dot{x}_1 = x_1, \qquad \dot{x}_2 = x_2$$

is orbitally unstable.

(b)   Verify that the system

$$\dot{x}_1 = x_2, \qquad \dot{x}_2 = 0$$

is orbitally stable but is unstable in the Liapunov sense.

4.   Show that the stability of the solution $\underline{x}^*$ of a general non-homogeneous equation

$$\dot{\underline{x}} = \underline{\underline{A}}\,\underline{x} + \underline{f}(t)$$

is determined by the zero solution of the homogeneous equation

$$\dot{\underline{z}} = \underline{\underline{A}}\,\underline{z}$$

Rewrite the second order equation

$$\ddot{x} + x = \cos t$$

as a set of two first order equations. Use the results obtained earlier to deduce that the solution is stable.

Solve the second order equation explicitly subject to the following initial conditions

(a) $x(0) = 1$, $\dot{x}(0) = 0$;   (b) $x^*(0) = 1 + \varepsilon_1$, $\dot{x}^*(0) = \varepsilon_2$

where $|\varepsilon_1| \ll 1$, $|\varepsilon_2| \ll 1$.

Show that the solution is unbounded but nonetheless stable.

5.   Determine if the following quadratic expressions are positive definite, semi-definite, or negative definite

(a) $V = x_1^2 + x_2^2 + x_3^2 - (1/2)\, x_1 x_2 - (1/2)\, x_1 x_3$

(b) $V = x_1^2 + x_2^2 + x_3^2 - 2\, x_1 x_2 + 4\, x_1 x_3 - 4\, x_2 x_3$

(c) $V = x_1^2 - x_2^2 + x_3^2 + 4\, x_1 x_2 - 2\, x_1 x_3$

Write V as $z^\dagger \underline{A}\, z$. Calculate the eigenvalues of $\underline{A}$ in all three cases. Can you discover any relationship between the sign of the eigenvalues of $\underline{A}$ and the positive definiteness of V?

6.   The Lorenz equations (see Chapter 5) can be written as

$$\dot{x}_1 = p\,(-x_1 + x_2)$$

$$\dot{x}_2 = r\,x_1 - (x_2 + x_1 x_3)$$

$$\dot{x}_3 = -b\,x_3 + x_1 x_2$$

where p, r, and b are positive constants.

Discuss the stability of the origin.

<u>Hint</u>: Consider two possible Liapunov functions: (a) $V = x_1^2 + p\,(x_2^2 + x_3^2)$

$$\text{(b)}\ \ V = (1/2)\,[\,r\,x_1^2 + p\,x_2^2 + p\,(x_3 - 2r)^2\,]$$

7.   By choosing an appropriate quadratic function V, examine the stability of the equilibrium points of the following systems

(a) $\dot{x}_1 = -x_1 - 2x_2^2, \qquad \dot{x}_2 = x_1 x_2 - x_2^3$

(b) $\ddot{x} + \dot{x}\, \sin x - x = 0$

(c) $\ddot{x} - \mu\dot{x}\,(x^2 - 1) + x = 0$ (Van der Pol equation)

<u>Hint</u>:   The equation in (c) is of the form of Equation (1.4-31) and it is useful to introduce Liénard coordinates. Define $F(x)$ as in Equation (1.4-32).

It follows that

$$\frac{dF}{dt} = \frac{dF}{dx}\frac{dx}{dt} = \dot{x}\, f(x)$$

Define

$$x_1 = x, \qquad x_2 = \dot{x}_1 + F(x), \qquad \dot{x}_2 = \ddot{x}_1 + \dot{x}_1 f(x_1) = -g(x_1)$$

Verify that, in this case, the equations are

$$\dot{x}_1 = x_2 + \mu\,[(1/3)\,x_1^3 - x_1]$$

$$\dot{x}_2 = -x_1$$

8.   Wu and Xiong (1998) have examined the stability of the solution of a fourth order differential equation which can be written as

$$\dot{x}_1 = x_2, \qquad \dot{x}_2 = x_3, \qquad \dot{x}_3 = x_4, \qquad \dot{x}_4 = -a_1 x_4 - a_2 x_3 - a_3 x_2 - f(x_1)$$

The Liapunov function V is chosen to be

$$2V = (x_4 + a_1 x_3 + \Delta x_2)^2 + (2/a_1^2) \int_0^{x_2} A(x_1)\, x_2\, dx_2$$

$$+ (a_3/a_1)\,[x_3 + a_1 x_2 + (a_1/a_3) f(x_1)]^2 + (2/a_1 a_3) \int_0^{x_1} f(x_1)\,A(x_1)\,dx_1$$

$$\Delta = a_2 - (a_3/a_1), \qquad A(x_1) = a_1 a_2 a_3 - a_3^2 - a_1^2\, f'(x_1)$$

Determine the conditions on f and a_i $(i = 1, 2, 3)$ for the system to be asymptotically stable.

9.   Euler's equations for a body spinning freely about a fixed point (in the absence of forces) are (Goldstein, 1972)

$$I_1 \dot{\omega}_1 - (I_3 - I_2)\,\omega_2 \omega_3 = 0$$

$$I_2 \dot{\omega}_2 - (I_1 - I_3)\,\omega_1 \omega_3 = 0$$

$$I_3 \dot{\omega}_3 - (I_2 - I_1)\,\omega_1 \omega_2 = 0$$

where I_1 , I_2 , and I_3 are the principal moments of inertia and $\underline{\omega}$ $(\omega_1, \omega_2, \omega_3)$ is the spin of the body along the principal axis fixed in the body.

Show that $(\omega_0, 0, 0)$ is a possible steady state solution and examine the stability of this solution by considering a perturbed state $(\omega_0 + \omega_1, \omega_2, \omega_3)$. A suitable Liapunov function is

$$V = I_2 (I_1 - I_2) \omega_2^2 + I_3 (I_1 - I_3) \omega_3^2 + [I_2 \omega_2^2 + I_3 \omega_3^2 + I_1 (\omega_1^2 + 2\omega_1 \omega_0)]^2$$

10.   Use the variable gradient method to generate a Liapunov function for the following systems and examine their stability

(a) $\dot{x}_1 = x_1 (x_2 - \alpha), \qquad \dot{x}_2 = x_2 (x_1 - \beta)$

    where α and β are constants

(b) $\dot{x}_1 = x_2, \qquad \dot{x}_2 = -(x_1^3 + x_2)$

11.   The linear equation that governs the motion of a gravitationally stabilized satellite in its pitch plane can be written as (Hsu and Meyer, 1968)

$$I_1 \ddot{\theta} + 3\Omega^2 (I_2 - I_1) \theta = u(t)$$

Here, I_1, I_2, and I_3 are the moments of inertia of the satellite about the pitch, roll and yaw axes respectively, Ω is the orbital angular rate, θ is the pitch plane angular deviation from the gravitational vertical, and u is the control that dampens the pitch oscillation.

Assume that u can be expressed as

$$u = -f(z), \qquad z = \alpha\theta + \beta\dot{\theta}$$

where α and β are constants.

The function f satisfies

$$f(0) = 0, \qquad z f(z) > 0, \qquad z \neq 0$$

Discuss the stability of the system, with and without control.

<u>Hint</u>: Choose $V = \frac{1}{2} (\theta^2 + a\dot{\theta}^2) + b \int_0^z f(\zeta) \, d\zeta, \quad a > 0, \ b > 0$

CHAPTER 4

BIFURCATION AND CATASTROPHE

4.1 INTRODUCTION

It is often assumed that a small change in input results in a small change in output, that is, the output is a continuous function of the input. This is not always true. Consider the process of heating a kettle of water. Near the boiling point, a small increase of heat could result in a change of state, from liquid to vapor, and this is a qualitative change. Bifurcation theory is the study of the point at which the qualitative behavior of a system changes.

We have mentioned examples of bifurcation in Chapters 1 and 3 and we now reconsider the Taylor stability problem examined in Chapter 3. There, we deduced that if the angular velocity Ω is less than a critical value Ω_c, the flow is laminar and the rotational flow is the only one which is observed. If $\Omega > \Omega_c$, the flow is unstable and Taylor vortices appear and the flow is three-dimensional. The rotational speed Ω can be considered to be a parameter and as we increase the value of Ω, we reach a critical value Ω_c at which a qualitative change occurs in the flow. This critical value is a point of bifurcation.

In this chapter, we consider a system that depends on a set of parameters which are denoted by $(\mu_1, \mu_2, ...)$. We assume the system to be autonomous and the set of equations describing the system can be written as

$$\frac{d\underline{x}}{dt} = \underline{f}(\underline{x}, \underline{\mu}) \tag{4.1-1}$$

where $\underline{x}$ and $\underline{\mu}$ are column vectors with components x_i and μ_i respectively.

Our objective is to determine the equilibrium states and their stability as the values of $\underline{\mu}$ are changed. If at some values of $\underline{\mu}$ ($= \underline{\mu}_0$), there is a qualitative change in the solution, then $\underline{\mu}_0$ are the bifurcation points. We recall that the equilibrium states are determined by solving

$$\underline{f}(\underline{x}, \underline{\mu}) = \underline{0} \tag{4.1-2}$$

In many systems, $\underline{f}$ can be expressed as the gradient of a scalar quantity. For these **gradient systems**, Equation (4.1-1) can be written as

$$\frac{d\underline{x}}{dt} = -\operatorname{grad}\phi\,(\underline{x},\,\underline{\mu}) \qquad\qquad\qquad (4.1\text{-}3)$$

The equilibrium states are given by

$$\operatorname{grad}\phi = \underline{0} \qquad\qquad\qquad (4.1\text{-}4)$$

Equation (4.1-4) states that the equilibrium states correspond to the stationary values of ϕ and the solution of Equation (4.1-4) describes a surface in the $(\underline{x},\,\underline{\mu})$ space. If the surface is smooth, a small change in $\underline{\mu}$ leads to a small change in $\underline{x}$ and nothing dramatic happens. However, if the surface is folded, then at the fold, a small change in $\underline{\mu}$ can result in a jump in the value of $\underline{x}$ and this is exactly the bifurcation point.

We recall that if $\phi(x)$ is a function of one variable, the stationary points are given by

$$\frac{d\phi}{dx} = 0 \qquad\qquad\qquad (4.1\text{-}5)$$

The stationary point is a maximum or minimum depending on the sign of the second derivative. Similarly, if $\phi\,(\underline{x},\,\underline{\mu})$ is a function of several variables, the stationary values are given by Equation (4.1-4). To determine if the stationary point is a maximum or minimum, we need to examine the sign of the determinants whose elements are the second derivative of ϕ (Chan Man Fong et al., 1997). If all the determinants vanish, no conclusion can be drawn and the point is a **singularity**.

The study of singularities is also known as **catastrophe theory**. Soon after the publication of Thom's book (1972) on catastrophe theory, this theory became extremely fashionable and was applied to a wide range of fields. Catastrophe theory is not a panacea to all problems and by the end of the 70's, a more sober view emerged. This theory can be used to describe qualitatively many phenomena and several applications are considered by Zeeman (1976). The understanding of the foundation of catastrophe theory demands a knowledge of topology which we do not assume the readers to possess. We shall not prove any theorem and we shall illustrate the applications by considering a few examples.

4.2   EXAMPLES OF BIFURCATION IN ONE DIMENSION

Example 4.2-1.   Pitchfork bifurcation

Determine the equilibrium points of the system

$$\frac{dx}{dt} = (\mu - \mu_c)\,x - a\,x^3 \qquad\qquad\qquad (4.2\text{-}1)$$

where a, μ, and μ_c are positive constants.

Examine the stability of the equilibrium points.

Equation (4.2-1) arises in the Taylor stability problem where μ can be associated with the Reynolds number and x with the amplitude of the Taylor vortices (Davey, 1962).

The three equilibrium points are

$$x_1 = 0, \qquad x_{2,3} = \pm \sqrt{(\mu - \mu_c)/a} \qquad\qquad (4.2\text{-}2a,b,c)$$

Note that if $\mu < \mu_c$, there is only one real solution, x_2 and x_3 are imaginary and can be ignored.

We now examine the stability of the equilibrium point $x = 0$. The linearized equation is

$$\frac{dx}{dt} = (\mu - \mu_c) x \qquad\qquad (4.2\text{-}3)$$

The solution of Equation (4.2-3) is

$$x = A_0 \exp[(\mu - \mu_c) t] \qquad\qquad (4.2\text{-}4)$$

where A_0 is a constant.

From Equation (4.2-4), we deduce that if $(\mu > \mu_c)$, $|x| \longrightarrow \infty$ as $t \longrightarrow \infty$ and the equilibrium point is unstable; if $(\mu < \mu_c)$, the equilibrium point is stable.

We next examine the equilibrium point x_2 $(= \sqrt{(\mu - \mu_c)/a})$. As in Chapter 1, we make a transformation and write

$$\bar{x} = x - x_2 \qquad\qquad (4.2\text{-}5)$$

Substituting Equation (4.2-5) into Equation (4.2-1) yields

$$\frac{d\bar{x}}{dt} = -[a\bar{x}^3 + 3ax_2\bar{x}^2 + 2ax_2^2\bar{x}] \qquad\qquad (4.2\text{-}6a)$$

$$\approx -2a[(\mu - \mu_c)/a]\bar{x} \qquad\qquad (4.2\text{-}6b)$$

$$\approx -2(\mu - \mu_c)\bar{x} \qquad\qquad (4.2\text{-}6c)$$

The equilibrium point x_2 exists only if $\mu > \mu_c$ and from Equation (4.2-6c), we deduce that $\bar{x}$ is stable. Similarly, we can show that x_3 is a stable equilibrium point.

We summarize our results as follows. If $\mu < \mu_c$, there is only one solution $(x = 0)$, that is, the Taylor vortices do not exist and the flow is purely rotational (laminar flow). If $\mu > \mu_c$, we have three possible solutions, the solution $x = 0$ is unstable and is not observed. The other two non-zero solutions x_2 and x_3 are stable and Taylor vortices with finite amplitude are observed. In Figure 4.2-1, we have drawn x as a function of μ.

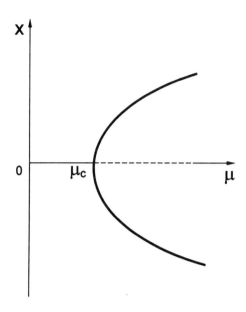

FIGURE 4.2-1   Pitchfork bifurcation
— : stable solution,   - - - : unstable solution

For $\mu < \mu_c$, $x = 0$ is the only solution, for $\mu > \mu_c$, three solutions are possible and the dotted lines indicate the unstable solution. At $\mu = \mu_c$, there is an exchange of stability, the solution $x = 0$ changes from a stable solution to an unstable solution. The point $\mu = \mu_c$ is a **pitchfork bifurcation** point as Figure 4.2-1 suggests. A pitchfork bifurcation is also known as a symmetric bifurcation. In the present example, two of the bifurcating equilibrium points are stable and this bifurcation is **supercritical**; otherwise, it is **subcritical**.

●

We note that for non-linear systems, we have more than one possible solution and not all the solutions are stable. Only the stable solutions are usually observed. A solution that is stable for $\mu < \mu_c$ can become unstable for $\mu > \mu_c$ as demonstrated in Example 4.2-1.

Example 4.2-2. **Transcritical bifurcation**

Discuss the stability of the equilibrium points of

$$\frac{dx}{dt} = \mu x - x^2 \tag{4.2-7}$$

where μ is a constant.

Equation (4.2-7) is a form of the logistic equation [Equation (1.3-39)]. The equilibrium points are

$$x = 0, \qquad x = \mu \qquad\qquad (4.2\text{-}8a,b)$$

If $\mu = 0$, the two solutions coalesce.

The solution $x = 0$ is stable if $\mu < 0$ and unstable if $\mu > 0$. To examine the stability of the solution $x = \mu$, we write

$$\bar{x} = x - \mu \qquad\qquad (4.2\text{-}9)$$

Substituting Equation (4.2-9) into Equation (4.2-7) yields

$$\frac{d\bar{x}}{dt} = -(\mu \bar{x} + \bar{x}^2) \qquad\qquad (4.2\text{-}10a)$$

$$\approx -\mu \bar{x} \qquad\qquad (4.2\text{-}10b)$$

From Equation (4.2-10b), we deduce that the solution $x = \mu$ is stable if $\mu > 0$ and unstable if $\mu < 0$.

We conclude that if $\mu < 0$, the solution $x = 0$ is stable and the solution $x = \mu$ is unstable; if $\mu > 0$, the solution $x = 0$ is unstable and the solution $x = \mu$ is stable. Figure 4.2-2 illustrates this result. The point $\mu = 0$ is a bifurcation point and is a **transcritical** (asymmetric) point.

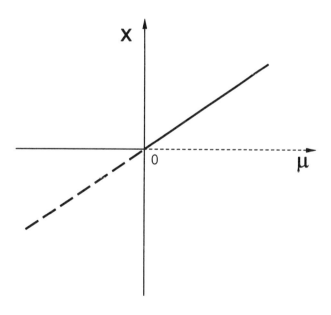

FIGURE 4.2-2   Transcritical bifurcation point
— : stable solution,   - - - : unstable solution

Example 4.2-3.   **Saddlenode bifurcation**

Examine the stability of the equilibrium points of

$$\frac{dx}{dt} = \mu + x^2 \tag{4.2-11a}$$

where μ is a parameter.

Equation (4.2-11a) has no equilibrium point if $\mu > 0$. For convenience, we rewrite Equation (4.2-11a) as

$$\frac{dx}{dt} = x^2 - \lambda^2 \tag{4.2-11b}$$

where $\mu = -\lambda^2$.

The equilibrium points are

$$x_1 = \lambda, \qquad x_2 = -\lambda \tag{4.2-12a,b}$$

We transfer the origin to λ by writing

$$\overline{x} = x - \lambda \tag{4.2-13}$$

Equation (4.2-11b) becomes

$$\frac{d\overline{x}}{dt} = 2\lambda\overline{x} + \overline{x}^2 \tag{4.2-14a}$$

$$\approx 2\lambda\overline{x} \tag{4.2-14b}$$

From Equation (4.2-14b), we deduce that the solution $x_1 = \lambda$ is unstable. Similarly, it can be shown that the solution $x_2 = -\lambda$ is stable. Equation (4.2-11a) has no equilibrium solution if $\mu > 0$ and two solutions if $\mu < 0$. The positive solution is unstable and the negative solution is stable. This is shown in Figure 4.2-3. The point $\mu = 0$ is a **saddlenode bifurcation** point.

●

We now consider the general one-dimensional problem involving one parameter. Equation (4.1-1) can be written as

$$\frac{dx}{dt} = f_{\prime}(x, \mu) \tag{4.2-15}$$

Suppose that (x_0, μ_0) corresponds to an equilibrium point and by definition

$$f(x_0, \mu_0) = 0 \tag{4.2-16}$$

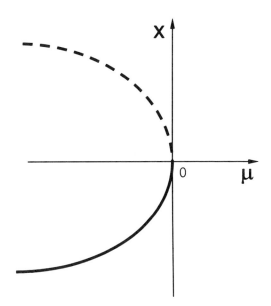

FIGURE 4.2-3   **Saddlenode bifurcation**
— : stable solution, - - - : unstable solution

To examine the stability of the equilibrium point, we expand $f(x, \mu)$ about (x_0, μ_0) and we obtain

$$f(x, \mu) = (x - x_0) \frac{\partial f}{\partial x} + (\mu - \mu_0) \frac{\partial f}{\partial \mu} + \frac{1}{2} (x - x_0)^2 \frac{\partial^2 f}{\partial x^2} + (x - x_0)(\mu - \mu_0) \frac{\partial^2 f}{\partial x \, \partial \mu}$$

$$+ \frac{1}{2} (\mu - \mu_0)^2 \frac{\partial^2 f}{\partial \mu^2} + \dots \tag{4.2-17a}$$

$$= a(\mu) + (x - x_0) \, b(\mu) + (x - x_0)^2 \, c(\mu) + \dots \tag{4.2-17b}$$

where $a(\mu) = (\mu - \mu_0) \dfrac{\partial f}{\partial \mu} + \dfrac{1}{2} (\mu - \mu_0)^2 \dfrac{\partial^2 f}{\partial \mu^2} + \dots$ (4.2-17c)

$$b(\mu) = (\mu - \mu_0) \frac{\partial^2 f}{\partial \mu \, \partial x} + \frac{1}{2} (\mu - \mu_0)^2 \frac{\partial^3 f}{\partial \mu^2 \, \partial x} + \dots \tag{4.2-17d}$$

$$c(\mu) = \frac{1}{2} \frac{\partial^2 f}{\partial x^2} + \frac{1}{2} (\mu - \mu_0) \frac{\partial^3 f}{\partial x^2 \, \partial \mu} + \dots \tag{4.2-17e}$$

and all the derivatives are evaluated at (x_0, μ_0).

If $\partial f/\partial x$ does not vanish at the equilibrium point, the system is **hyperbolic**; otherwise, it is **non-hyperbolic**. Usually, bifurcation is associated with non-hyperbolic equilibrium points and in Equation (4.2-17d), we have assumed that the system is non-hyperbolic. The study of bifurcation is essentially the determination of the zeros of

$$a(\mu) + (x - x_0) \, b(\mu) + (x - x_0)^2 \, c(\mu) = 0 \tag{4.2-18}$$

Glendenning (1994) has shown that the bifurcation point is

(a)   pitckfork if $\dfrac{\partial f}{\partial \mu} = 0$, $\dfrac{\partial^2 f}{\partial x^2} = 0$, $\dfrac{\partial^2 f}{\partial \mu \, \partial x} \neq 0$, $\dfrac{\partial^3 f}{\partial x^3} \neq 0$;   (4.2-19a-d)

(b)   transcritical if $\dfrac{\partial f}{\partial \mu} = 0$, $\dfrac{\partial^2 f}{\partial x^2} \neq 0$, $\left(\dfrac{\partial^2 f}{\partial \mu \, \partial x}\right)^2 > \dfrac{\partial^2 f}{\partial x^2} \dfrac{\partial^2 f}{\partial \mu^2}$;   (4.2-20a-c)

(c)   saddlenode if $\dfrac{\partial f}{\partial \mu} \neq 0$, $\dfrac{\partial^2 f}{\partial x^2} \neq 0$.   (4.2-21a,b)

We can verify that Examples 4.2-1 to 3 satisfy the criteria (a) to (c) respectively. We illustrate the application of the criteria stated earlier in the next example.

Example 4.2-4. **Proportional harvesting**

Determine the type of bifurcation point(s) of

$$\frac{dx}{dt} = \mu x - b x^2 - h x \tag{4.2-22}$$

where b, h, and μ are positive constants.

Equation (4.2-22) represents a population which grows according to the logistic equation [Equation (1.3-39)] and is harvested at a rate proportional to the size of the population.

The equilibrium points are

$$x_1 = 0, \qquad x_2 = (\mu - h)/b \tag{4.2-23a,b}$$

The population has to be positive, so x_2 exists only if $\mu > h$.

In this example, the function f is

$$f = (\mu - h) x - b x^2 \tag{4.2-24}$$

Differentiating and evaluating the derivatives at x_1 yields

$$\frac{\partial f}{\partial \mu} = 0, \quad \frac{\partial^2 f}{\partial x^2} = -b \neq 0, \quad \left(\frac{\partial^2 f}{\partial \mu \, \partial x}\right)^2 = 1, \quad \frac{\partial^2 f}{\partial \mu^2} = 0 \qquad \text{(4.2-25a-e)}$$

From Equations (4.2-20a-c, 25a-e), we deduce that the bifurcation point is a transcritical point. From Figure 4.2-2, we conclude that if $\mu < h$, the population will become extinct ($x = 0$ is the only stable solution) and if $\mu > h$, the population will tend to $(\mu - h)/b$ which is the stable solution. This result is to be expected. If the rate of harvesting is greater than the rate of growth, then the population will be exterminated.

In this example, we have considered μ to be the only variable, b and h have fixed values. If μ, b, and h are variable parameters, the equilibrium points describe a surface. If there are two variable parameters, the equilibrium points describe a curve. We do not consider these cases, instead we consider the two-dimensional problems

4.3   TWO-DIMENSIONAL PROBLEMS

In this section, we present a few examples of bifurcations involving two dependent variables (x_1, x_2) or second order differential equations.

Example 4.3-1.   **Perturbed harmonic oscillator**

Examine qualitatively the solution of

$$\frac{d^2 x}{dt^2} - \mu \frac{dx}{dt} + x = 0 \qquad \text{(4.3-1)}$$

If $\mu < 0$, Equation (4.3-1) is the equation of a damped oscillator.

By letting $x = x_1$, Equation (4.3-1) can be written as

$$\begin{bmatrix} \dot{x}_1 \\ \\ \dot{x}_2 \end{bmatrix} = \begin{bmatrix} 0 & 1 \\ \\ -1 & -\mu \end{bmatrix} \begin{bmatrix} x_1 \\ \\ x_2 \end{bmatrix} \qquad \text{(4.3-2)}$$

The origin $(0, 0)$ is the equilibrium point. From Chapter 1, section 2, we deduce that the origin is a stable focus if $\mu < 0$, an unstable focus if $\mu > 0$, and a stable center if $\mu = 0$. The bifurcation point is $\mu = 0$.

To draw a bifurcation diagram, we associate the periodic solution $(\mu = 0)$ with a non-zero amplitude a. In the case of the stable focus $(\mu < 0)$, the amplitude a is zero and when $\mu > 0$, the solution is unstable. The bifurcation diagram is shown in Figure 4.3-1 and this type of bifurcation is the **vertical bifurcation**.

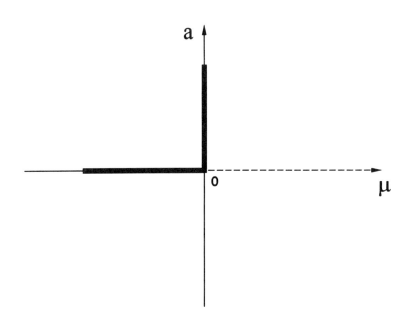

FIGURE 4.3-1   Vertical bifurcation
— : **stable solution,   - - - : unstable solution**

Example 4.3-2.   Hopf bifurcation

Discuss the solution of

$$\dot{x}_1 = x_2 + x_1 (\mu - x_1^2 - x_2^2)$$ (4.3-3a)

$$\dot{x}_2 = -x_1 + x_2 (\mu - x_1^2 - x_2^2)$$ (4.3-3b)

where μ is a parameter.

The origin is an equilibrium point.

To solve Equations (4.3-3a, b), we introduce polar coordinates (r, θ) as defined by Equations (1.2-28a, b). Following the procedure adopted in Example 1.3-1, Equations (4.3-3a, b) become

$$\dot{r} = r (\lambda^2 - r^2)$$ (4.3-4a)

$$\dot{\theta} = -1 \qquad (4.3\text{-}4b)$$

where $\lambda^2 = \mu$.

The solutions of Equations (4.3-4a, b) are respectively

$$r^2 = \mu / (1 + c_0 e^{-2\mu t}) \qquad (4.3\text{-}5a)$$

$$\dot{\theta} = -t + c_1 \qquad (4.3\text{-}5b)$$

where c_0 and c_1 are constants.

From Equation (4.3-5a), we deduce that if $\mu < 0$, $r \longrightarrow 0$ as $t \longrightarrow \infty$, that is, the trajectories spiral towards the equilibrium point. If $\mu > 0$, the trajectories spiral towards the circle of radius $\sqrt{\mu}$, that is, the equilibrium point is unstable (r does not tend to 0) but $|r|$ does not tend to infinity, it tends to $\sqrt{\mu}$. Thus, the unstable solution tends to a periodic solution. The trajectories of the present system is similar to those shown in Figure 1.3-1.

The bifurcation point is $\mu = 0$ and if we associate the periodic solution with an amplitude a, the bifurcation diagram is as shown in Figure 4.3-2. This is **Hopf bifurcation**.

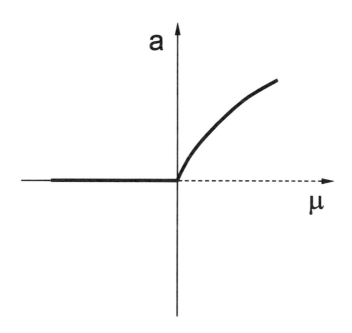

FIGURE 4.3-2   Hopf bifurcation
— : stable solution,   - - - : unstable solution

The two examples have shown that the structure and variety of bifurcation phenomena in two dimensions are much richer than in one dimension, in particular the unstable solution might tend to a periodic solution. Hopf bifurcation is useful in many applications and we shall state Hopf's theorem without proof.

Theorem 1

Suppose that the system

$$\dot{x}_1 = f_1(x_1, x_2, \mu) \tag{4.3-6a}$$

$$\dot{x}_2 = f_2(x_1, x_2, \mu) \tag{4.3-6b}$$

has an equilibrium state at $(0, 0, \mu_0)$.

The Jacobian matrix

$$\underline{\underline{J}} = \begin{bmatrix} \dfrac{\partial f_1}{\partial x_1} & \dfrac{\partial f_1}{\partial x_2} \\[2ex] \dfrac{\partial f_2}{\partial x_1} & \dfrac{\partial f_2}{\partial x_2} \end{bmatrix} \tag{4.3-7}$$

evaluated at $(0, 0)$ has eigenvalues $\alpha(\mu) \pm i\beta(\mu)$ with $\alpha(\mu_0) = 0$, $\beta(\mu_0) \neq 0$.

Further

$$\left. \frac{d\alpha}{d\mu} \right|_{\mu_0} \neq 0 \tag{4.3-8}$$

then, in the neighborhood of $(0, 0, \mu_0)$, there is a non-trivial periodic solution.

In the next example, we apply this theorem to the Brusselator (Example 1.3-7).

Example 4.3-3. **Brusselator**

Show that the Brusselator has a periodic solution.

The equations of the Brusselator are given by Equations (1.3-81a, b). To be constant with our present notation, we replace β by μ. We assume the value of α is fixed and we denote it by a. Equations (1.3-81a, b) can be written as

$$\frac{dx_1}{dt} = 1 - (1 + \mu) x_1 + a x_1^2 x_2 \tag{1.3-81a}$$

$$\frac{dx_2}{dt} = \mu x_1 - a x_1^2 x_2 \tag{1.3-81b}$$

The equilibrium point is $(1, \mu/a, \mu_0)$.

For convenience, we transform the origin to $(1, \mu/a)$ and write

$$x_1^* = x_1 - 1 \tag{4.3-9a}$$

$$x_2^* = x_2 - \mu/a \tag{4.3-9b}$$

Equations (1.3-81a, b) become

$$\dot{x}_1^* = x_1^*(\mu - 1) + a x_2^* + 2 a x_1^* x_2^* + x_1^{*2}(a x_2^* + \mu) \tag{4.3-10a}$$

$$\dot{x}_2^* = -\mu x_1^* - a x_2^* - 2 a x_1^* x_2^* - x_1^{*2}(\mu + a x_2^*) \tag{4.3-10b}$$

The Jacobian $\underline{J}$ evaluated at $(0, 0)$ is

$$\underline{J} = \begin{bmatrix} \mu - 1 & a \\ -\mu & -a \end{bmatrix} \tag{4.3-11}$$

The eigenvalues are

$$\alpha + i\beta = \frac{1}{2}\left[(1 + a - \mu) \pm \sqrt{(1 + a - \mu)^2 - 4a}\,\right] \tag{4.3-12}$$

The eigenvalues are purely imaginary if

$$\alpha = 0 \quad \Rightarrow \quad \mu = \mu_0 = (1 + a) \tag{4.3-13a,b,c}$$

From Equation (4.3-12), we deduce that

$$\frac{d\alpha}{d\mu} = -\frac{1}{2} \neq 0 \tag{4.3-14a,b}$$

All the conditions of Theorem 1 are satisfied and we conclude that in the neighborhood of

$$x_1 = 1, \qquad x_2 = \mu/a, \qquad \mu = 1 + a \tag{4.3-15a,b,c}$$

there is a periodic solution.

Note that the bifurcation point given by Equation (4.3-13) is identical to that given in Example 1.3-7.

●

The condition that the eigenvalues of $\underline{\underline{J}}$ are purely imaginary implies that the linear system is a periodic solution and from Table 1.3-1, no definite conclusion can be drawn for the non-linear system. However, condition (4.3-8) allows us to conclude that there is a periodic solution in the neighborhood of the equilibrium point.

It seems that Theorem 1 was known to Poincaré who thought it was too trivial to write it down (Glendinning, 1994). The theorem was also proved independently by Andronov and some authors refer to Theorem 1 as the Poincaré-Andronov-Hopf theorem. A proof and a brief history of Theorem 1 are given in Hale and Koçak (1991).

We next apply Theorem 1 to confirm the result of Example 1.4-5. There, it was shown, by using Poincaré-Bendixson theorem, that the Van der Pol equation has a periodic solution.

Example 4.3-4. **Van der Pol equation**

Show that the Van der Pol equation has a periodic solution.

Equations (1.4-26b, c), which are reproduced here, are the equations of Van der Pol

$$\dot{x}_1 = x_2 \tag{1.4-26b}$$

$$\dot{x}_2 = -x_1 + \mu(1 - x_1^2)x_2 \tag{1.4-26c}$$

The equilibrium state is $(0, 0, \mu_0)$.

The Jacobian $\underline{\underline{J}}$ is

$$\underline{\underline{J}} = \begin{bmatrix} 0 & 1 \\ -1 & \mu \end{bmatrix} \tag{4.3-16}$$

The eigenvalues of $\underline{\underline{J}}$ are

$$\alpha + i\beta = \frac{1}{2}\left[-2\mu \pm \sqrt{4\mu^2 - 4}\,\right] \tag{4.3-17}$$

The eigenvalues are purely imaginary if

$$\mu = \mu_0 = 0 \tag{4.3-18a,b}$$

The bifurcation point is $(0, 0, 0)$.

From Equation (4.3-17), we obtain

$$\frac{d\alpha}{d\mu} = -1 \neq 0 \qquad\qquad (4.3\text{-}19a,b)$$

All the conditions of Theorem 1 are satisfied and the set of Equations (1.4-26b, c) has a periodic solution, as deduced in Example 1.4-5.

●

We have considered only continuous systems. In the next section, we examine discrete systems.

4.4   DISCRETE SYSTEMS

In a continuous model, the independent variable is assumed to change continuously and the dependent variable generally changes continuously. In many situations, changes do not occur continuously. For example, the population of a species changes only as a result of a birth or a death and both events do not occur continuously. For many simple organisms, the time lapse between generations is very short and a continuous model is a reasonable approximations. For other cases, a discrete model is more appropriate.

In a discrete model, we relate the value of the independent variable v at time $(t + 1)$ to the value of v at time t. This is expressed mathematically as

$$v_{t+1} = f(v_t) \qquad\qquad (4.4\text{-}1)$$

where f is usually a continuous function of v_t.

Equation (4.4-1) describes a **first order discrete dynamical system**. If f is given, Equation (4.4-1) can be solved recursively for every given initial condition.

The **equilibrium value** of v (v_e) is the value of v such that

$$v_t = v_e \qquad\qquad (4.4\text{-}2)$$

for all values of t.

Combining Equations (4.4-1, 2), we deduce that v_e is obtained by solving

$$v_e = f(v_e) \qquad\qquad (4.4\text{-}3)$$

The equilibrium value is **stable** if there is a number $\varepsilon > 0$ such that if v_0 is the initial value of v (at $t = 0$)   and

$$|v_0 - v_e| < \varepsilon \qquad\qquad (4.4\text{-}4)$$

then

$$v_t \longrightarrow v_e \quad \text{as} \quad t \longrightarrow \infty \qquad\qquad (4.4\text{-}5a,b)$$

The equilibrium value is **unstable** if statement (4.4-4) holds and

$$|v_t - v_e| > \varepsilon \qquad\qquad (4.4\text{-}6)$$

for some values of t and not necessarily for all values of t greater than a critical value t_c.

To determine the conditions on f for stability, we proceed in the usual manner. We perturb the equilibrium state and write

$$v_t = v_e + w_t, \quad |w_t| \ll 1 \qquad\qquad (4.4\text{-}7)$$

Substituting Equation (4.4-7) into Equation (4.4-1) yields

$$v_e + w_{t+1} = f(v_e + w_t) \qquad\qquad (4.4\text{-}8)$$

Expanding the right side of Equation (4.4-8) and using Equation (4.4-3), we obtain

$$w_{t+1} = w_t \, f'(v_e) + O(w_t^2) \qquad\qquad (4.4\text{-}9a)$$

$$\approx \lambda \, w_t \qquad\qquad (4.4\text{-}9b)$$

where $\lambda \; [= f'(v_e)]$ is a constant.

The solution of Equation (4.4-9b) is

$$w_t = \lambda^t \, w_0 \qquad\qquad (4.4\text{-}10)$$

where w_0 is the initial value of w.

From Equation (4.4-10), we deduce that as $t \longrightarrow \infty$

$$w_t \longrightarrow 0, \qquad \text{if } |\lambda| < 1 \qquad\qquad (4.4\text{-}11a)$$

$$|w_t| \longrightarrow \infty, \quad \text{if } |\lambda| > 1 \qquad\qquad (4.4\text{-}11b)$$

This implies that the equilibrium point is stable if $|f'(v_e)| < 1$ and is unstable if $|f'(v_e)| > 1$. No conclusion can be drawn if $|f'(v_e)| = 1$ and we have to consider higher derivatives.

If $f'(v_e) = 1$, $f''(v_e) < 0$, v_e is semistable from above, that is, if $v_0 > v_e$, $v_t \longrightarrow v_e$ as $t \longrightarrow \infty$, but if $v_0 < v_e$, v_t does not tend to v_e as $t \longrightarrow \infty$. If $f''(v_e) > 0$, v_e is semistable from below. In this case, if $v_0 < v_e$, $v_t \longrightarrow v_e$ as $t \longrightarrow \infty$, but if $v_0 > v_e$, v_t does not tend to v_e as $t \longrightarrow \infty$. If $f'(v_e) = 1$, $f''(v_e) = 0$, and $f'''(v_e) \neq 0$, v_e is stable if $f'''(v_e) < 0$ and unstable if $f'''(v_e) > 0$.

If $f'(v_e) = -1$, we compute

$$g'''(v_e) = -2f'''(v_e) - 3[f''(v_e)]^2 \tag{4.4-12}$$

If $g'''(v_e) < 0$, v_e is stable and if $g'''(v_e) > 0$, v_e is unstable. A proof of these criteria is given in Sandefur (1990).

We illustrate the concepts introduced in this section by some examples.

Example 4.4-1. **Logistic equation**

Write down the discrete logistic equation, find the equilibrium values, and examine their stability.

The continuous logistic equation is given by Equation (1.3-39). Let N_t denote the population at time t and the logistic equation is

$$N_{t+1} - N_t = a N_t (1 - N_t/K) \tag{4.4-13}$$

where a and K are positive constants.

Equation (4.4-13) can also be written as

$$N_{t+1} = (1 + a) N_t - a N_t^2/K \tag{4.4-14}$$

Note that if Equation (4.4-14) is linearized, it reduces to Equation (4.4-9b) and N_t tends to infinity as t tends to infinity.

We non-dimensionalize Equation (4.4-14) by writing

$$N_t = K v_t \tag{4.4-15}$$

Equation (4.4-14) becomes

$$v_{t+1} = (1 + a) v_t - a v_t^2 \tag{4.4-16}$$

The equilibrium points are obtained by solving

$$v_e = (1 + a) v_e - a v_e^2 \tag{4.4-17}$$

The solutions are

$$v_e = 0, \qquad v_e = 1 \tag{4.4-18a,b}$$

From Equation (4.4-1, 16), we deduce that

$$f'(v_t) = (1 + a) - 2\,a\,v_t \tag{4.4-19}$$

It follows that

$$f'(0) = 1 + a, \qquad f'(1) = 1 - a \tag{4.4-20a,b}$$

Since a is positive, we deduce that the origin $(v_e = 0)$ is unstable and the other equilibrium point $(v_e = 1)$ is stable if $0 < a < 2$. If $a = 2$, the equilibrium point $(v_e = 1)$ is semistable $[f''(1) = -4]$.

●

In Example 4.4-1, we have seen that the number of equilibrium points is independent of the parameter a. Suppose that we have a system that depends on a parameter μ and that we introduce this parameter explicitly in Equation (4.4-1). The equilibrium points are given by

$$v_e = f(v_e, \mu) \tag{4.4-21}$$

The number of equilibrium points N_e may depend on μ. If N_e changes as μ is increased from $\mu_e - \varepsilon$ to $\mu_e + \varepsilon$, for some positive ε, μ_e is a **bifurcation point**.

We next consider a few examples of bifurcation.

Example 4.4-2. **Fixed harvesting**

Determine the equilibrium points of

$$v_{t+1} = 1.8\,v_t - 0.8\,v_t^2 - \mu \tag{4.4-22}$$

and examine their stability.

Equation (4.4-22) models a population that grows according to the logistic equation (a = 0.8) and μ units of the species are harvested per unit time.

The equilibrium values are given by

$$v_e = 1.8\,v_e - 0.8\,v_e^2 - \mu \tag{4.4-23}$$

The solutions are

$$v_e = 0.5\,(1 \pm \sqrt{1 - 5\mu}\,) \tag{4.4-24}$$

The quantity v_e must be real and there are two equilibrium points if $\mu < 0.2$. If $\mu = 0.2$, there is only one equilibrium point and if $\mu > 0.2$, there is no equilibrium point. N_e is given by

$$N_e = \begin{cases} 2, & \text{if } \mu < 0.2 \\ 1, & \text{if } \mu = 0.2 \\ 0, & \text{if } \mu > 0.2 \end{cases} \qquad (4.4\text{-}25)$$

The bifurcation point is $\mu = 0.2$.

If $\mu > 0.2$, v_t will eventually be negative implying that the species has become extinct due to overharvesting.

Form Equation (4.4-22), we can compute f' and $f'(v_e)$ is given by

$$f'(v_e) = 1.8 - 1.6\, v_e \qquad (4.4\text{-}26)$$

Substituting the values of v_e into Equation (4.4-25), we deduce that the point $0.5\,(1 + \sqrt{1 - 5\mu})$ is stable and the other equilibrium point is unstable.

The bifurcation diagram is the plot of v_e versus μ and is shown in Figure 4.4-1. Note that Figures 4.2-3 and 4.4-1 are similar. This type of bifurcation is the saddlenode bifurcation.

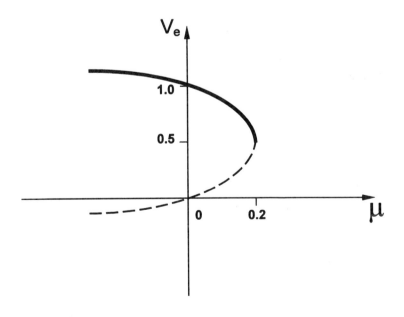

FIGURE 4.4-1.   Bifurcation diagram for fixed harvest
— : stable solution, - - - : unstable solution

Example 4.4-3. **Cubic dependence**

Show that $\mu = 0$ is a bifurcation point of

$$v_{t+1} = (1+\mu)\, v_t - v_t^3 \tag{4.4-27}$$

The equilibrium points are

$$v_e = 0, \qquad v_e = \pm\sqrt{\mu} \tag{4.4-28a,b,c}$$

If μ is negative, there is only one equilibrium point and if μ is positive, there are three equilibrium points.

We next examine the stability of these equilibrium points. Computing f' yields

$$f'(0) = 1+\mu, \qquad f'(\pm\sqrt{\mu}\,) = 1-2\mu^2 \tag{2.2-29a,b}$$

The equilibrium point $(v_e = 0)$ is unstable if $\mu > 0$ or $\mu < -2$, and is stable if $-2 < \mu < 0$. If $\mu = 0$, the equilibrium point is stable $[f'''(0) = -6]$, and if $\mu = -2$, the equilibrium point is unstable $[g'''(0) = 12]$. Note that on increasing μ from -2, the equilibrium point changes from a stable state to an unstable state on crossing $\mu = 0$.

The equilibrium points $(v_e = \pm\sqrt{\mu}\,)$ are stable if $0 < \mu < 1$.

The bifurcation diagram is similar to Figure 4.2-1 with $\mu_c = 0$. The point $\mu = 0$ is a bifurcation point and this type of bifurcation is known as pitchfork bifurcation.

Example 4.4-4. **Proportional harvesting**

Determine the bifurcation point of

$$v_{t+1} = 1.5\, v_t - 0.5\, v_t^2 - \mu v_t \tag{4.4-30}$$

Equation (4.4-30) is the discrete analog of Equation (4.2-22). It represents a population growth according to the logistic equation [Equation (4.4-17), $a = 0.5$] and a rate of harvest proportional to the size of the population.

The equilibrium points are given by

$$v_e = 1.5\, v_e - 0.5\, v_e^2 - \mu v_e \tag{4.4-31}$$

The solutions are

$$v_e = 0, \qquad v_e = 1 - 2\mu \tag{4.4-32a,b}$$

From Equations (4.4-32a, b), we deduce that if $\mu < 1/2$, there are two equilibrium points. If $\mu = 1/2$, the two equilibrium points coalesce. If $\mu > 1/2$, $1 - 2\mu$ is negative and is not a possible solution.

To examine the stability of the equilibrium points, we calculate f' from Equation (4.4-30). Evaluating f' at the two equilibrium points, we obtain

$$f'(0) = 1.5 - \mu, \qquad f'(1 - 2\mu) = 0.5 + \mu \qquad\qquad (4.4\text{-}33a,b)$$

From Equation (4.4-33a), we deduce that the origin ($v_e = 0$) is unstable of $\mu < 0.5$ or $\mu > 2.5$. It is stable if $2.5 > \mu > 0.5$. If $\mu = 0.5$, the origin is semistable $[f''(0) = -1]$.

The second equilibrium point given by Equation (4.4-33b) is stable of $-1.5 < \mu < 0.5$ and is unstable for other values of μ. When $\mu = 0.5$, the equilibrium point is semistable.

We can summarize the results as follows. As μ is increased from 0, it is found that on crossing $\mu = 0.5$, the number of equilibrium points changes form 2 to 1. There is also an interchange of stability, the equilibrium point ($v_e = 0$) changes form being unstable to being stable and the other point ($v_e = 1 - 2\mu$) changes in the reverse direction, that is, if $\mu \geq 0.5$, the only solution is $v_e = 0$. Figure 4.4-2 illustrates this situation. The point $\mu = 0.5$ is a bifurcation point and is a transcritical bifurcation. Note the similarity between Figures 4.2-2 and 4.4-2.

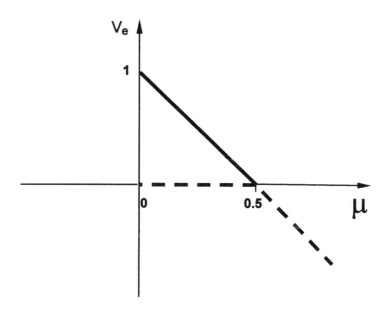

FIGURE 4.4-2.   Transcritical bifurcation
— : stable solution, - - - : unstable solution

●

May (1976) has shown that simple difference equations can exhibit complicated dynamic behavior. He has also suggested possible areas of application.

By confining our attention to one-dimensional problems, we have been able to illustrate the results in a plane. For higher order equations, more complex types of bifurcations arise. In Section 4.3, we have considered one two-dimensional case for a continuous system. We do not consider higher order equations for the discrete system and we refer interested readers to Sandefur (1990) for discussions on higher order recurrence equations. In the next section, we shall discuss catastrophe.

4.5   ELEMENTARY CATASTROPHE

In Chapter 3, we have seen that the stability properties of a system can be deduced by examining the properties of a Liapunov function. In this section, we briefly discuss the possible description of bifurcation properties of a system via a function. For example, consider the function

$$\phi = \frac{1}{3} x^3 - \mu x \qquad (4.5\text{-}1)$$

If ϕ represents the potential of a gradient system, the equilibrium points of the system are given by (Equation 4.1-4)

$$\phi' = x^2 - \mu = 0 \qquad (4.5\text{-}2a,b)$$

The solution is

$$x_0 = \pm \sqrt{\mu} \qquad (4.5\text{-}3)$$

There are two equilibrium points if $\mu > 0$, one if $\mu = 0$, and none if $\mu < 0$. This situation is similar to the situation in Examples 4.2-1, 3 and 4.4-2, 3. To determine if the equilibrium points correspond to a maximum or minimum, we have to consider the second derivative

$$\phi'' = 2x \qquad (4.5\text{-}4)$$

The negative (positive) root corresponds to a maximum (minimum) and if $\mu = 0$, we have the indeterminate case. This indeterminate situation can be associated with a bifurcation point.

We now consider a function of one independent variable $\phi(x)$.

The **singular points** of $\phi(x)$ are given by

$$\phi'(x) = 0 \qquad (4.5\text{-}5)$$

Without loss of generality, we can assume that

$$\phi(0) \ = \ \phi'(0) \ = \ 0 \tag{4.5-6a,b}$$

This implies that the origin is a singular point. We expand $\phi(x)$ in a Taylor series about the origin. The leading terms are

$$\phi(x) \ = \ \frac{x^2}{2}\,\phi''(0) + \frac{x^3}{3!}\,\phi'''(0) \ + \ ... \tag{4.5-7}$$

The indeterminate case corresponds to $\phi''(0) = 0$, and the leading term is x^3. If $\phi''(0) = 0$, the origin is a **degenerate singular point** and the **degree of degeneracy** of ϕ is 2.

If ϕ is a function of more than one variable, the situation is more complex. For simplicity, we consider the case of two variables. The critical point of $\phi(x_1, x_2)$ is obtained by solving simultaneously

$$\frac{\partial\phi}{\partial x_1} \ = \ 0, \qquad \frac{\partial\phi}{\partial x_2} \ = \ 0 \tag{4.5-8a,b}$$

To determine if the critical point is a maximum or minimum, we introduce the **Hessian** H defined by

$$H \ = \ \begin{vmatrix} \dfrac{\partial^2\phi}{\partial x_1^2} & \dfrac{\partial^2\phi}{\partial x_1\partial x_2} \\[2ex] \dfrac{\partial^2\phi}{\partial x_1\partial x_2} & \dfrac{\partial^2\phi}{\partial x_2^2} \end{vmatrix} \tag{4.5-9}$$

The condition for a maximum or a minimum is $H > 0$; the critical point corresponds to a maximum if $\partial^2\phi/\partial x_1^2 < 0$ and to a minimum if $\partial^2\phi/\partial x_1^2 > 0$. If $H < 0$, the critical point corresponds to a saddle point. A degenerate point corresponds to $H = 0$, and to examine the nature of the critical points, we need to consider the higher derivatives. However, unlike the case of one variable, $H = 0$ does not imply that all three second order partial derivatives are zero. From Equation (4.5-9), H vanishes if

$$\frac{\partial^2\phi}{\partial x_1^2}\frac{\partial^2\phi}{\partial x_2^2} \ = \ \left(\frac{\partial^2\phi}{\partial x_1\partial x_2}\right)^2 \tag{4.5-10}$$

If we expand ϕ about the origin, even in the degenerate case, we need to include the quadratic terms, unless all the second order partial derivatives are zero. In general, ϕ can be expanded as

$$\phi \ = \ \frac{1}{2}\,(a\,x_1^2 + 2b\,x_1 x_2 + c\,x_2^2) + \text{higher order terms} \tag{4.5-11a}$$

where

$$a = \partial^2 \phi / \partial x_1^2, \qquad b = \partial^2 \phi / \partial x_1 \partial x_2, \qquad c = \partial^2 \phi / \partial x_2^2 \qquad (4.5\text{-}11b,c,d)$$

$$ac = b^2 \qquad (4.5\text{-}11e)$$

Equation (4.5-11e) implies that a and c have the same sign and the quadratic expression in Equation (4.5-11a) is a perfect square. Equation (4.5-11a) can be written as

$$\phi = \pm \frac{1}{2} (\alpha x_1 + \beta x_2)^2 + \dots \qquad (4.5\text{-}12)$$

where $\alpha = \sqrt{|a|}$, $\beta = \sqrt{|b|}$; we have to take the positive sign if a is positive and the negative sign if a is negative.

Equation (4.5-12) suggests a rotation of the axes and we write

$$u_1 = (\alpha x_1 + \beta x_2) / \sqrt{\alpha^2 + \beta^2}, \qquad u_2 = (\beta x_1 - \alpha x_2) / \sqrt{\alpha^2 + \beta^2} \qquad (4.5\text{-}13a,b)$$

Using the chain rule, we obtain

$$\frac{\partial \phi}{\partial u_1} = \frac{\partial \phi}{\partial u_2} = \frac{\partial^2 \phi}{\partial u_2^2} = \frac{\partial^2 \phi}{\partial u_1 \partial u_2} = 0 \qquad (4.5\text{-}14a\text{-}d)$$

$$\frac{\partial^2 \phi}{\partial u_1^2} = \pm (\alpha^2 + \beta^2) \qquad (4.5\text{-}14e)$$

where all the derivatives are evaluated at the singular point.

In the u_1-direction, ϕ has a maximum or minimum depending on the sign of Equation (4.5-14e). We verify if this is true for all u_i-directions. Let

$$v = u_1 \sin \theta + u_2 \cos \theta \qquad (4.5\text{-}15)$$

Evaluating the derivatives at the singular point yields

$$\frac{d\phi}{dv} = \frac{\partial \phi'}{\partial u_1} \sin \theta + \frac{\partial \phi}{\partial u_2} \cos \theta = 0 \qquad (4.5\text{-}16a,b)$$

$$\frac{d^2\phi}{dv^2} = \sin^2 \theta \frac{\partial^2 \phi}{\partial u_1^2} + 2 \sin \theta \cos \theta \frac{\partial^2 \phi}{\partial u_1 \partial u_2} + \cos^2 \theta \frac{\partial^2 \phi}{\partial u_2^2} = \sin^2 \theta \frac{\partial^2 \phi}{\partial u_1^2} \qquad (4.5\text{-}16c,d)$$

In the $\theta = 0$ direction (u_2-direction), the Taylor series expansion reduces to

$$\phi = \frac{u_2^3}{3!} \frac{\partial^3 \phi}{\partial u_2^3} + \dots \tag{4.5-17}$$

This means that we are back to the single variable case and can ignore u_1. This result can be extended to the case of n variables. The Hessian matrix is

$$\underline{\underline{H}} = \begin{bmatrix} \dfrac{\partial^2 \phi}{\partial x_1^2} & \cdots & \dfrac{\partial^2 \phi}{\partial x_1 \partial x_n} \\ \vdots & & \\ \dfrac{\partial^2 \phi}{\partial x_n \partial x_1} & \cdots & \dfrac{\partial^2 \phi}{\partial x_n^2} \end{bmatrix} \tag{4.5-18}$$

If the rank of $\underline{\underline{H}}$ is $n - r$ for some $r > 0$ (that is, $|H| = 0$), there is a coordinate transformation that allows us to express ϕ as

$$\phi = \pm x_{r+1}^2 \pm \dots \pm x_n^2 + g(x_1, \dots, x_r) \tag{4.5-19}$$

where the leading terms in g are cubic.

For our purpose, we need to consider only $x_1, \dots, x_r$ and can ignore the variables $x_{r+1}, \dots, x_n$. In catastrophe theory, we need not consider all the n variables, we need to consider only the r variables $(x_1, \dots, x_r)$ and the number r is the **corank** of $\underline{\underline{H}}$. The number r can be interpreted as the number of directions in which the function is degenerate.

In Equation (4.4-1), we have introduced a set of parameters μ_i and if we have k such parameters, k is the **codimension** (codim). The number of types of degenerate functions depends on the corank and codimension of the system. It has been shown that if the codimension is less than or equal to four, the number of types of singularity is seven and these seven types are **Thom's seven elementary catastrophes**. Table 4.5-1 lists the seven elementary catastrophes.

A complete derivation of Table 4.5-1 is given in Zeeman (1977) and we do not reproduce it here. Instead, we give heuristic arguments to show the plausibility of the fold catastrophe. We have seen earlier that by expanding the function $\phi(x)$ in a Taylor series, the leading term is x^3 if the origin is a degenerate singular point. ϕ can be approximated as

$$\phi = x^3 \tag{4.5-20}$$

TABLE 4.5-1

The seven elementary catastrophes

Corank	Codim	Singularity	Name	Universal unfolding (potential)
1	1	x^3	fold	$x^3 + \mu_1 x$
1	2	x^4	cusp	$x^4 + \mu_1 x^2 + \mu_2 x$
1	3	x^5	swallowtail	$x^5 + \mu_1 x^3 + \mu_2 x^2 + \mu_3 x$
1	4	x^6	butterfly	$x^6 + \mu_1 x^4 + \mu_2 x^3 + \mu_3 x + \mu_4 x$
2	3	$x_1^3 + x_2^3$	hyperbolic umbilic	$x_1^3 + x_2^3 + \mu_1 x_1 x_2 + \mu_2 x_1 + \mu_3 x_2$
2	3	$x_1^3 - 3 x_1 x_2^2$	elliptic umbilic	$x_1^3 - 3 x_1 x_2^2 + \mu_1 (x_1^2 + x_2^2) + \mu_2 x_1 + \mu_3 x_2$
2	4	$x_1^2 x_2 + x_2^4$	parabolic umbilic	$x_1^2 x_2 + x_2^4 + \mu_1 x_1^2 + \mu_2 x_2^2 + \mu_3 x_1 + \mu_4 x_2$

We now examine if it possesses all the singular properties associated with cubic functions. The singular points are

$$\phi' = 3 x^2 = 0 \qquad\qquad (4.5\text{-}21a,b)$$

The two singular points are coincident. We perturb the function slightly and let ϕ_1 be

$$\phi_1 = x^3 - \mu x \qquad\qquad (4.5\text{-}22)$$

Equation (4.5-22) is similar to Equation (4.5-1) and in this case, there are two equilibrium points which are not coincident. We have stated earlier that if $\mu < 0$ there is no equilibrium point, if $\mu = 0$ there is one equilibrium point, and if $\mu > 0$ there are two equilibrium points. On perturbing $\phi = x^3$ slightly, we obtain results which are qualitatively different and the function ϕ defined by Equation (4.5-20) is **structurally unstable**. We next perturb ϕ_1 and we write ϕ_2 as

$$\phi_2 = x^3 - \mu x + a \qquad\qquad (4.5\text{-}23)$$

The equilibrium points are

$$\phi_2' = 3x^2 - \mu = 0 \tag{4.5-24a,b}$$

From Equation (4.5-24b), we deduce that ϕ_1 and ϕ_2 share the same properties, that is, two equilibrium points if $\mu > 0$, one equilibrium point if $\mu = 0$, and zero equilibrium point if $\mu < 0$. The function ϕ_1 is **structurally stable**. Next, we consider ϕ_3 defined by

$$\phi_3 = x^3 + bx_2 - \mu x + a \tag{4.5-25}$$

We change the origin by writing

$$x' = x + b/3 \tag{4.5-26}$$

Substituting Equation (4.5-26) into Equation (4.5-25) yields

$$\phi_3 = x'^3 - (\mu + b/3)x' + (a - \mu b/3 - b^3/27) \tag{4.5-27}$$

By a change of origin, ϕ_3 can be written as ϕ_2, and ϕ_2 has the same singular properties as ϕ_1. This means that for our purpose all cubic expressions can be written as ϕ_1, and ϕ_1 is the **universal unfolding** of x^3. Similar heuristic arguments can be used to justify the other catastrophes (Saunders, 1980).

We next discuss the geometry of the catastrophes.

4.6 GEOMETRY OF ELEMENTARY CATASTROPHES

In this section, we examine the geometrical properties of some catastrophes. We denote the universal unfolding (potential) by ϕ. The equilibrium surface M is defined by

$$\text{grad } \phi = 0 \tag{4.6-1}$$

This surface M is generated by the critical points of ϕ. The degenerate points are given by

$$H = 0 \tag{4.6-2}$$

where H is the Hessian.

The set of degenerate critical points is denoted by S. The **bifurcation set** B is obtained by eliminating the state variables ($\underline{x}$) from Equations (4.6-1, 2). We now consider the fold.

(a)   The Fold

The potential ϕ for the fold is given in Table 4.5-1 and can be written as

$$\phi = x^3 + \mu_1 x \qquad (4.6\text{-}3)$$

The equilibrium surface M is given by

$$\phi' = 3x^2 + \mu_1 = 0 \qquad (4.6\text{-}4a,b)$$

The set S is defined by

$$\phi'' = 6x = 0 \qquad (4.6\text{-}5a,b)$$

Eliminating x from Equations (4.6-4b, 5b), we obtain B which is given by

$$\mu_1 = 0 \qquad (4.6\text{-}6)$$

That is to say, the bifurcation set is only one point ($\mu_1 = 0$). The bifurcation set divides the control space ($\underline{\mu}$-space) into two regions. If $\mu_1 > 0$, Equation (4.6-4b) has no real root and ϕ has no critical point. If $\mu_1 < 0$, ϕ has two critical points, one is a maximum (unstable) and the other a minimum (stable). If $\mu_1 = 0$, the critical point is a point of inflexion. We can now conclude that if a system has only one parameter and ϕ is a suitable potential then the system has a stable solution only if μ_1 is negative. The equilibrium surface M is of the shape shown in Figure 4.2-3.

(b)   The Cusp

For the cusp catastrophe, ϕ is given by (see Table 4.5-1)

$$\phi = x^4 + \mu_1 x^2 + \mu_2 x \qquad (4.6\text{-}7)$$

The critical points are given by

$$\phi' = 4x^3 + 2\mu_1 x + \mu_2 = 0 \qquad (4.6\text{-}8)$$

Equation (4.6-8) is cubic and the number of real roots can be one or three. If $8\mu_1^3 + 27\mu_2^2 \leq 0$, we have three real roots, otherwise we have only one real root.

The degenerate critical points are given by

$$\phi'' = 12x^2 + 2\mu_1 = 0 \qquad (4.6\text{-}9)$$

Eliminating x between Equations (4.6-8, 9) yields the set B which is given by

$$8\mu_1^3 + 27\mu_2^2 = 0 \qquad (4.6\text{-}10)$$

Note the relationship between the condition for the existence of three or one real root and the bifurcation set [Equation (4.6-10)].

The equilibrium surface is the folded surface and is shown in Figure 4.6-1. The projection of the folded part of this surface determines the bifurcation set B, as shown in Figure 4.6-2.

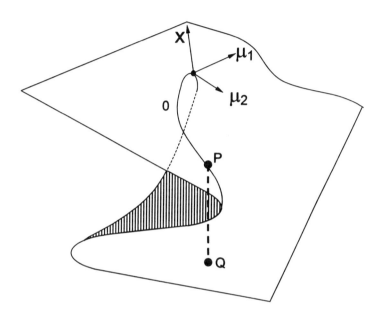

FIGURE 4.6-1   The equilibrium surface of the cusp

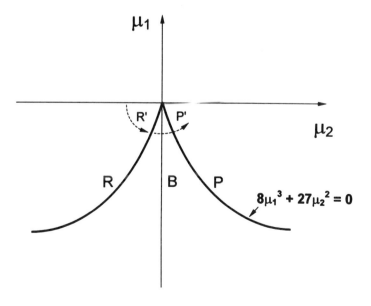

FIGURE 4.6-2   Bifurcation set of the cusp

For all values of the control parameters (μ_1, μ_2) not lying in B, Equation (4.6-8) has one real root and from Equation (4.6-9), we deduce that ϕ'' is positive and this state is stable. For values of μ_1 and μ_2 lying in B, there are three roots. One root corresponds to a maximum (unstable state) and two roots to a minimum (stable state). This means that in the folded region of the equilibrium surface (Figure 4.6-1), the upper and lower surfaces represent the stable (observable) states and the fold part represents the unstable (not observable) state. We now consider the trajectory of a system on the upper surface. On reaching the fold, point P in Figure 4.6-1, the system jumps from point P to a point Q on the lower surface because the pleat part is unstable. This can be better illustrated by considering a cross-section of the equilibrium surface. Figure 4.6-3 shows a cross-section of the surface for a constant value of μ_1 (< 0).

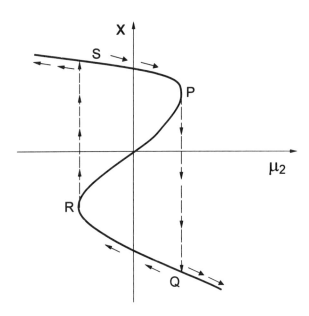

**FIGURE 4.6-3   Section of the equilibrium surface of a cusp.
Arrows indicate direction of trajectory**

As explained earlier, when the trajectory of the system reaches the point P, the system jumps to point Q on the lower surface. The system does not follow the path PR, because PR corresponds to the unstable state. However, if initially the system is on the lower surface and as the control parameter μ_2 is decreased, the trajectory follows the path QR. At R, the system jumps to point S, on the upper surface because, as mentioned earlier, RP is an unstable state. As the system moves from the upper to the lower and back to the upper surface, the system exhibits jump discontinuities and hysteresis.

In the bifurcation diagram (Figure 4.6-2), we have marked the points R and P, the points at which jump discontinuities occur for μ_1 = constant. If we allow μ_1 and μ_2 to vary, we can draw the trajectory $(R' \, P')$ and if the direction is from left to right (as indicated by the arrows), the jump discontinuity occurs at P'. If the trajectory is in the opposite direction, the jump discontinuity occurs at R'.

Note that if μ_1 is positive, jump discontinuities cannot occur. The variable x is a continuous function of μ_1 and μ_2. As the values of μ_1 are decreased to negative values, the possibility of discontinuity arises. For this reason, μ_1 is known as the **splitting factor** and μ_2 as the **normal factor**.

The other catastrophes involving three or more control variables are more difficult to analyze explicitly. For example, the phase space of the swallowtail catastrophe is a four-dimensional space (three control parameters and one state variable) and we are unable to represent it geometrically. However, it is possible to synthesize the phase space by considering various cross-sections : say, $\mu_1 = $ constant. For a detailed analysis of the remaining five elementary catastrophes, we refer the readers to Thom (1972), Woodcock and Poston (1974), and Saunders (1980). In the next section, we consider the applications of catastrophe theory.

4.7 APPLICATIONS

Catastrophe theory has been applied to physical and social sciences and we give a few examples here.

Example 4.7-1. **Euler arch**

The Euler arch consists of two light rigid rods of unit length, each with one end free and the other end joined by a pivot and a spring of modulus G, as shown in Figure 4.7-1. A weight of magnitude F_2 is applied at the pivot and a horizontal compressive force F_1 acts at each of the three ends. The angle that either rod makes with the horizontal is denoted by θ. If $F_1 = F_2 = 0$, the rods are horizontal, the angle between them is π and $\theta = 0$. If F_1 and F_2 are non-zero, $\theta \neq 0$ and the angle between the rods is $\pi - 2\theta$.

We now calculate the energy of the system. The energy of the spring is $\frac{1}{2} G (2\theta)^2$, the energy of the load is $F_2 |1| \sin \theta$ (the rods are of unit length), the work done by the compressive force is $2F_1 (1 - \cos \theta)$. The total energy is

$$\phi = 2G\theta^2 + F_2 \sin \theta - 2F_1 (1 - \cos \theta) \qquad (4.7\text{-}1)$$

The equilibrium surface M is

$$\phi' = 4G\theta + F_2 \cos \theta - 2F_1 \sin \theta = 0 \qquad (4.7\text{-}2a,b)$$

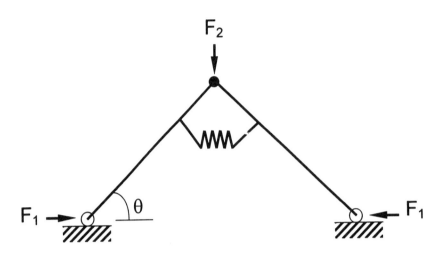

FIGURE 4.7-1   Euler arch

The degenerate singular points are given by

$$\phi'' = 4G - F_2 \sin\theta - 2F_1 \cos\theta = 0 \qquad (4.7\text{-}3a,b)$$

We consider the case of $F_2 = 0$. Equation (4.7-2b) simplifies to

$$2G\theta = F_1 \sin\theta \qquad (4.7\text{-}4)$$

Equation (4.7-4) has only one root ($\theta = 0$) if $2G > F_1$ and from Equation (4.7-3b), we deduce that ϕ'' is positive and the equilibrium point $\theta = 0$ is stable. If $F_1 > 2G$, Equation (4.7-4) has three roots ($\theta = 0$, $\theta = \pm\theta_0$; $2G\theta_0 = F_1 \sin\theta_0$) and the equilibrium point $\theta = 0$ is unstable and the other two are stable. We conclude that if $F_1 < 2G$, buckling does not occur and the two rods remain horizontal. Buckling occurs only if $F_1 > 2G$, but the direction of buckling is not predicted.

We now examine the case $F_2 \neq 0$. It is not possible to solve Equation (4.7-2b) analytically. We assume that θ is small and expand $\sin\theta$ and $\cos\theta$ in powers of θ. The function ϕ can be approximated as

$$\phi \approx F_2\theta + (2G - F_1)\,\theta^2 - (1/6)\,F_2\theta^3 + (1/12)\,F_1\theta^4 \qquad (4.7\text{-}5)$$

We can write ϕ in the canonical form by introducing a new variable x defined by

$$x = \theta - (F_2/2F_1) \qquad (4.7\text{-}6)$$

We also assume that F_2 is small, its square and higher powers can be neglected and $F_1 \approx 2G$, that is, we consider the case of near buckling as discussed earlier. Equation (4.7-5) now becomes

$$\phi \approx \frac{1}{6} G x^4 + \mu_1 x^2 + F_2 x \tag{4.7-7}$$

where $\mu_1 = 2G - F_1$ is small.

The function ϕ is now written in the standard form of the cusp catastrophe (Table 4.5-1). The equilibrium surface M is given by

$$x^3 + (3\mu_1/G) x + 3F_2/2G = 0 \tag{4.7-8}$$

The degenerate critical points are obtained by solving

$$x^2 + \mu_1/G = 0 \tag{4.7-9}$$

The bifurcation set B is

$$16\mu_1^3/G + 9F_2^2 = 0 \tag{4.7-10}$$

We can interpret our results as follows. Let $\mu_1 < 0$, $F_2 = 0$ and the rods are buckled upwards. We now apply F_2 downwards and increase its magnitude from zero until it reaches a critical value F_{2c} when the system snaps suddenly into the downwards equilibrium position. As discussed earlier F_{2c} corresponds to the point P in Figure 4.6-2 and the value of F_{2c} is determined from Equation (4.7-10) and is given by

$$F_{2c} = (4/3) (\mu_1^3/G)^{1/2} \tag{4.7-11}$$

This behavior has been observed in some structures. Thompson (1982) has examined more complex structures.

Note that from Equation (4.7-4), we have deduced that at $F_1 = 2G$, we have a pitchfork bifurcation (Figure 4.2-1). The similarity between Equations (4.2-1 and 7-4) can be brought out more clearly by expanding $\sin \theta$ in powers of θ up to and including θ^3. We can regard F_2 as an imperfection and via catastrophe theory, we have deduced the effect of an imperfection on the system past the bifurcation point.

Example 4.7-2. **Taylor stability**

In the discussion of the Taylor stability problem in Chapter 3 and Example 4.2-1, it is assumed that the length of the cylinders is infinite. In reality, the cylinders are of finite length. Benjamin (1978) has considered the case of finite length. In this problem, we have two control parameters, the Reynolds number Re and the length of the cylinder ℓ. From Examples 4.2-1 and 7-1, we are led to express the equilibrium surface as

$$x^3 - (Re - Re_c) \, x - \varepsilon = 0 \qquad\qquad (4.7\text{-}12)$$

where ε is associated with ℓ and is regarded as an imperfection of the ideal system ($\ell \longrightarrow \infty$). [Here, we have replaced μ in Equation (4.2-1) by Re.]

The bifurcation set is

$$4 \, (Re - Re_c)^3 = 27 \, \varepsilon^2 \qquad\qquad (4.7\text{-}13)$$

Inside the bifurcation set, we have three real roots and outside the bifurcation set, there is one real root. From Equation (4.7-13), we deduce that Taylor cells appear when $Re > R_1$ and R_1 is given by

$$R_1 = Re_c + (27 \, \varepsilon^2/4)^{1/3} \qquad\qquad (4.7\text{-}14)$$

Note that in the ideal case, Taylor cells appear abruptly at $Re = Re_c$ whereas in the imperfect case, the critical Reynolds number is a continuous function of ε. Since the imperfection cannot be quantified exactly, there is no precise critical Reynolds number. If the cylinders are very long ($\ell/d \gg 1$, where d is the annular gap between the cylinders), the development of cells is very rapid in the neighborhood of Re_c. Benjamin (1978) has noted that the appearance of the cells in short cylinders is a smooth process and has cited other authors who have made similar observations.

From the discussions on the results shown in Figure 4.6-3, we can expect that the present system exhibits hysteresis. Indeed, Benjamin (1978) has found the Reynolds number at which cells are established is higher if Re is increased from below to above R_1 than if Re is decreased from above to below R_1.

The experimental observations of Benjamin (1978) can be explained by the catastrophe theory.

Example 4.7-3. **Phase transition**

The process of boiling and condensation can be modeled as a cusp catastrophe. We assume that the gas obeys the Van der Waals equation, which can be written as

$$V^3 - (b + RT/P) \, V^2 + (a/P) \, V - ab/P = 0 \qquad\qquad (4.7\text{-}15)$$

where V is the volume, R is the gas constant, T is the temperature, P is the pressure, a and b are constants.

The state variable is V and the control parameters are P and T. Equation (4.7-15) is a cubic in V and can be regarded as the equilibrium surface of a cusp. As discussed earlier, if one of the control parameters (e.g. T) is varied, there is a possibility of a jump discontinuity in V and this discontinuity is observed at the condensation or boiling point. However, the boiling and condensation points are coincident and usually no hysteresis is observed. This is in disagreement with the predictions arrived at earlier.

To explain this discrepancy, two types of change of state have been proposed, namely the **perfect delay** and the **Maxwell conventions**. These conventions can be stated as follows.

(a)   *Perfect Delay Convention.* A system remains in the equilibrium state it is in, until that state disappears and then moves to the nearest local minimum. This means that the jump occurs when the trajectory crosses the bifurcating curves (points denoted by P' and R' in Figure 4.6-2).

(b)   *Maxwell Convention.* A system always seeks a global minimum. As soon as the local minimum ceases to be a global minimum, the system immediately moves to the new global minimum. In this case, the jump occurs along the μ_1-axis in Figure 4.6-2 and no hysteresis is observed.

Other conventions have been proposed and a discussion on this topic is given in Gilmore (1979).

In the present example, we have to adopt the Maxwell convention because the process is controlled by the laws of statistical mechanics and an averaging process is involved. Usually, if the noise of the system is of the same order of magnitude as the jump discontinuity, we adopt the Maxwell convention. It should also be pointed out that with sufficient care, it is possible to superheat a liquid or to supercool a vapor.

Example 4.7-4.   **Spruce budworm outbreak**

Casti (1982) has shown, via catastrophe theory, that outbreaks of spruce budworm are inevitable. This insect outbreak occurs irregularly at 40-80 years interval and has been the subject of various studies. The model adopted by Casti (1982) is the LJH (Ludwig, Jones, and Holling) model and is described by the following equations

$$\frac{dB}{dt} = \alpha_1 B \left[1 - B(\alpha_3 + E^2)/(\alpha_2 S E^2)\right] - \alpha_4 B^2/(\alpha_5 S^2 + B^2) \qquad (4.7\text{-}16a)$$

$$\frac{dS}{dt} = \alpha_6 S \left[1 - \alpha_7 S/(\alpha_8 E)\right] \qquad (4.7\text{-}16b)$$

$$\frac{dE}{dt} = \alpha_9 E (1 - E/\alpha_7) - \alpha_{10} B E^2/S (\alpha_3 + E^2) \qquad (4.7\text{-}16c)$$

where B is budworm density, S is the total surface area of the branches in the stand of tress, E represents the stand "reserve energy" which is a measure of the stand foliage condition and health of the trees. The parameters α_1 to α_{10} represent various intrinsic growth rates, predation rates, and so on.

We denote the non-zero equilibrium values of B, S, and E by B_e, S_e, and E_e respectively. Setting the right sides of Equations (4.7-16a-c) to zero and combining Equations (4.7-16a, b) yields

$$\alpha_1 \alpha_7^3 (\alpha_3 + E_e^2) B_e^3 - \alpha_1 \alpha_2 \alpha_7^2 \alpha_8 E_e^3 B_e^2 + (\alpha_1 \alpha_5 \alpha_7 \alpha_8^2)(\alpha_3 + E_e^2) B_e^3$$

$$+ \alpha_2 \alpha_4 \alpha_7^2 \alpha_8 E_e^3 B_e - \alpha_1 \alpha_2 \alpha_5 \alpha_8^3 E_e^5 = 0 \tag{4.7-17a}$$

Combining Equations (4.7-16b, c) yields

$$B_e = -\alpha_8 \alpha_9 (E_e^3 - \alpha_7 E_e^2 + \alpha_3 E_e + \alpha_3 \alpha_7) / \alpha_7^2 \alpha_{10} \tag{4.7-17b}$$

The quantity of interest is B_e and Equation (4.7-17a) can be considered to be the equilibrium surface with B_e as state variable and E_e as a parameter. It has been verified that for all ecologically possible values of α_i and for a given value of B_e, Equation (4.7-17b) has only one solution which is in the physically relevant range ($0 \le E_e \le 1$). Thus, it is justifiable to regard E_e as an additional parameter. Equation (4.7-17a) can be written in the canonical form by introducing a new variable x defined by

$$x = B_e - \alpha_2 \alpha_8 E_e^3 / [3\alpha_7 (\alpha_3 + E_e^2)] \tag{4.7-18}$$

Substituting Equation (4.7-18) into Equation (4.7-17a) yields

$$-(x^3 + \mu_1 x + \mu_2) = 0 \tag{4.7-19a}$$

$$\mu_1 = -\alpha_8^2 E_e^2 (\beta_1 - \beta_2 - \alpha_5 \alpha_8) / \alpha_7^2 \tag{4.7-19b}$$

$$\mu_2 = -\alpha_2 \alpha_8^2 E_e^5 (2\beta_1 - 3\beta_2 + 6\alpha_5 \alpha_8) / 9 \alpha_7^3 (\alpha_3 + E_e)^2 \tag{4.7-19c}$$

$$\beta_1 = \alpha_2^2 \alpha_8 E_e^4 / 3 (\alpha_3 + E_e)^2 \tag{4.7-19d}$$

$$\beta_2 = \alpha_2 \alpha_4 \alpha_7 E_e / \alpha_1 (\alpha_3 + E_e^2) \tag{4.7-19e}$$

The 11 parameters (α_1 to α_{10}, E_e) are combined into two control parameters μ_1 and μ_2. The bifurcation set is given by

$$4\mu_1^3 + 27\mu_2^2 = 0 \tag{4.7-20}$$

If $4\mu_1^3 + 27\mu_2^2 > 0$, Equation (4.7-19a) has only one real root and a stable population of spruce budworm is possible. From an intelligent guess of the values of the 11 parameters, it is found that $4\mu_1^3 + 27\mu_2^2 < 0$ and catastrophe is possible. As the values of α_i and E_e vary, the values of μ_1 and μ_2 vary and the possibility of a jump discontinuity in x is possible. The parameters of α_i and E_e must be adjusted so that x always lies in the lower surface in Figure 4.6-1. We need to ensure that the trajectory does not cross the curve $R'R$ in Figure 4.6-2. If the trajectory crosses $R'R$, then the value of x jumps to the upper surface of Figure 4.6-1 and we have an outbreak. When an outbreak occurs, the state of health of the forest worsens leading to a decrease in the value of E_e. This in turn leads to an increase in the values of μ_1 and μ_2. From Figure 4.6-2, it is seen that when the trajectory crosses the curve $P'P$, the value of x drops to the lower surface. Given that $4\mu_1^3 + 27\mu_2^2 < 0$, it seems that outbreaks are inevitable. This boom and bust phenomenon appears to be part of nature, unless external control is applied.

Casti (1982) has also examined the possibility of external control, such as application of insecticides.

Example 4.7-5. **Economic recession**

Varian (1979) examined Kaldor's model of business cycles using the method of catastrophe. The basic equations of the model are

$$\frac{dy}{dt} = s\,[C(y) + I(y, k) - y] \tag{4.7-21a}$$

$$\frac{dk}{dt} = I(y, k) - I_0 \tag{4.7-21b}$$

where y is the gross national income, k is the capital stock, $C(y)$ is the consumption function, $I(y, k)$ is the gross investment function, I_0 represents investment for replacing depreciated equipment, and s represents the speed of response of income adjustments.

A savings function $S(y)$ is defined by

$$S(y) = y - C(y) \tag{4.7-22}$$

It is assumed that $C(y)$ is a linear function and can be written as

$$C(y) = c_0 + c_1 y \tag{4.7-23}$$

where c_0 and c_1 are constants.

The investment function is a decreasing function of k and is a non-linear function of y. The shape of I as a function of y is sigmoidal as shown in Figure 4.7-2. This means that I grows logistically [Equation (1.3-39)] with y.

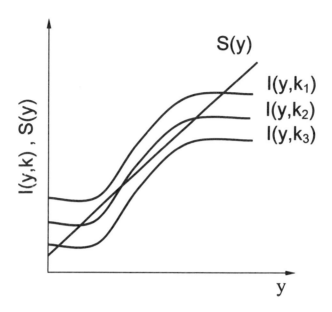

**FIGURE 4.7-2   Savings S (y) and investment I (y, k) functions
against income y for various values of capital stock k**

The equilibrium surface is obtained by setting the right side of Equation (4.7-21a) to zero and can be written as

$$I = S \qquad\qquad (4.7\text{-}24)$$

In this example, we have only one control variable (k) and this corresponds to the fold catastrophe.

From Figure 4.7-2, it is seen that for some values of k, Equation (4.7-24) has three roots and for other values of k, it has only one root. It can be shown that if there are three solutions, two of them are stable and one is unstable. If there is only one solution, it is a stable solution. The equilibrium curve is as shown in Figure 4.7-3.

As in Figure 4.6-3, the upper and lower parts of the curve correspond to the stable solution and the middle part corresponds to the unstable solution. Suppose we start from a point on the upper curve. As k increases, y decreases until a critical value of k is reached (point P in Figure 4.7-3) when y jumps down to the lower curve (point Q). The economy is now in recession, investment is encouraged and since I is a decreasing function of k, this results in a decrease in the value of k. The trajectory is now along QR and when it reaches R, y moves up to the upper curve and the recession is over. This process repeats itself resulting in a cycle.

To model both recessions and depressions, Varian (1979) introduced another parameter w associated with wealth. With two control parameters, we have the cusp catastrophe. With such a model, it is possible to explain both recessions and depressions. Using the results given in Chapter 1, Section 4, it can be shown that the Kaldor model has periodic solutions (Varian, 1979).

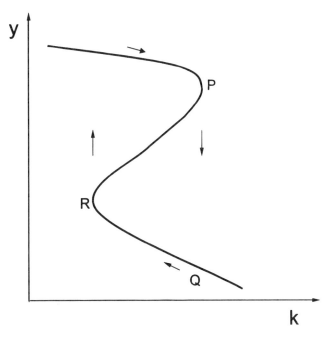

**FIGURE 4.7-3   Equilibrium curve on an economic model.
Arrows indicate the direction of trajectory**

Example 4.7-6.   **Aggression in dogs**

This example is one of the earliest application of catastrophe theory in animal behavior. Zeeman's model of aggressive behavior in dogs is based on Lorenz's observation that the two factors which determine aggression in dogs are fear and rage. A dog's rage is related to the degree to which its mouth is opened or its teeth are bared; its fear is measured by how much its ears are flattened. The magnitudes of this two control parameters (fear and rage) can be estimated. The behavior of the dog is the state variable and the various possible modes are described on the equilibrium surface such as the one shown in Figure 4.6-1. The upper surface represents a state of most likely to be aggressive and the lower surface represents the submissive mood. Near the center, where both rage and fear are nearly equal, both submissive and aggressive attitudes are possible. The neutral attitude is unstable and very unlikely.

Suppose the dog is in an aggressive mood and on the behavior (equilibrium) surface its attitude is represented by a point on the upper surface. If the dog's fear now begins to increase while its rage remains at the same level, its emotional state is still on the upper state until its trajectory reaches the edge of the pleat (point P in Figure 4.6-1). At this point, with a slight increase in fear, the mood changes suddenly and the dog becomes submissive and flees. In the same way, the mood can change suddenly from being submissive to being aggressive. For further details, see Zeeman (1976).

4.8   COMMENTS

Thom (1976) stated that "the range of Catastrophe Theory extends from pure mathematics to the most far fetched speculations in traditional philosophy." Examples 4.7-1 to 6 can be considered to be the illustration of this statement. Example 4.7-1 is an example of the theory's application in the usual physical sense, all quantities are defined clearly and are quantified. In Example 4.7-2, the parameter ε associated with imperfection is not quantified; although, we have interpreted ε as a measure of the finiteness of the length of the cylinders, it could also equally be used to describe the eccentricity. Thus ε may be associated with more than one physical idea. By the time we reach Example 4.7-6, we have stopped writing equations!

The applications of catastrophe theory in biological and social sciences have been controversial and have attracted a lot of criticisms (Sussmann and Zahler, 1987; Zahler and Sussmann, 1977). The origins and controversies of catastrophe theory are summarized in Woodcock and Davis (1978).

We end this chapter by noting that the catastrophe theory describes local and static behavior and is based on a variational principle. It cannot be used to predict situations far from equilibrium states and on a global scale. It provides an explanation for the frequently observed jump discontinuity in the output as the input is varied continuously.

PROBLEMS

1.   Find the equilibrium points, examine their stability, and discuss the bifurcation properties of the following systems

     (a) $\dot{x} = x\,(\mu - x^2)$

     (b) $\dot{x} = x\,(\mu - x - x^2)$

     (c) $\dot{x} = \mu x - bx^2 - h$ (this equation represents a model of constant harvesting   h,   see Example 4.4-2).

2.   Determine the equilibrium points and determine the types of bifurcations [use Equations (4.2-19 to 21)] of the following systems

     (a) $\dot{x} = x^3 + x^2 - (2 + \mu)\,x + \mu$

     (b) $\dot{x} = x\,(x^2 - \mu)\,(x^2 + \mu^2 - 1)$

3.   Deduce that if $\omega \neq 0$, the origin is the only equilibrium point of the equations

$$\dot{x}_1 = \mu x_1 - \omega x_2 - x_1 (x_1^2 + x_2^2)$$

$$\dot{x}_2 = \omega x_1 + \mu x_2 - x_2 (x_1^2 + x_2^2)$$

Does the system has a Hopf bifurcation at $(0, 0, 0)$?

Rewrite the differential equations in polar form and deduce the limit cycle.

4.   Show that the system

$$\ddot{x} - (\mu - x^2) \dot{x} + x = 0$$

has a Hopf bifurcation at $(0, 0, 0)$.

Use an appropriate theorem (see Chapter 1) to deduce that the system has a periodic solution.

5.   Consider the following discrete systems

(a) $v_{t+1} = v_t^2 - 3v_t + 3$

(b) $v_{t+1} = -0.5 v_t^3 + 0.5 v_t^2 + v_t$

Determine the equilibrium points and examine their stability.

6.   Find the equilibrium points of the discrete system

$$v_{t+1} = 3 v_t - v_t^2 - \mu$$

for various values of μ.

Examine the stability of the equilibrium points and draw the bifurcation diagram.

7.   Find the singular points, the degenerate and the bifurcation sets of the following functions

(a) $\phi = 2x^3 + \mu_1 x^2 + \mu_2 x$

(b) $\phi = x^5 + \mu_1 x^3 + \mu_2 x^2 + \mu_3 x$

(c) $\phi = x_1^3 + x_2^3 + \mu_1 x_1 x_2 + \mu_2 x_1 + \mu_3 x_2$

8.   The Duffing equation can be written as

$$\ddot{x} + k \dot{x} + x + a x^3 = F \cos (1 + \omega) t$$

where k, a, F, and ω are constants.

It is usual to assume that k, a, and ω are small, k is positive, and a can be positive (hard spring) or negative (soft spring).

Attempt a solution of the form

$$x = A \cos [(1 + \omega) t - \phi]$$

Deduce, after making the appropriate simplifications, that

$$\tan \phi = 4k / (3aA^2 - 8\omega)$$

$$A^2 \left(\frac{3}{4} aA^2 - 2\omega \right)^2 = F^2 - k^2 A^2$$

Note that we have a cubic in A^2 and this suggests a cusp catastrophe. Discuss the solution in the framework of catastrophe theory.

<u>Hint</u>: See Holmes and Rand (1976).

CHAPTER 5

CHAOS

5.1 INTRODUCTION

Chaos has aroused considerable interest among the general public and several popular books (Gleick, 1987; Stewart, 1989) are available. Chaos is often associated with randomness, but here we are dealing with deterministic systems and for such systems, we expect predictability and order. However, in recent years, we have discovered that the trajectories described by non-linear systems can be heavily dependent on the initial conditions. Two trajectories may start close to each other, they may diverge exponentially and yet are confined in a three-dimensional space as described in Chapter 1. After a finite time, the two trajectories can be completely different and this is referred to as **deterministic chaos**. The trajectories do not tend to infinity (being confined to a finite space) and they are attracted to a **strange attractor**.

The realization that even for a deterministic system our prediction may have a limited range has prompted Lighthill (1986) to apologize on behalf of the "broad global fraternity of practitioners of mechanics" for having misled the general educated public about the determinism of systems obeying Newton's laws of motion. As early as the end of the 19th century, Poincaré (1892) discovered that certain Hamiltonian systems can exhibit chaotic motion. But at the time, it was believed that this was a rarity. It was not until after the publication of Lorenz's (1963) paper that scientists started to study chaos in earnest. A brief history of chaos is given in Diacu and Holmes (1996). Several major papers on this topic are reproduced in Hao (1990).

Chaos has been observed in both conservative and dissipative systems and is believed to occur frequently. In the next section, we describe a few chaotic systems.

5.2 EXAMPLES

Example 5.2-1. Logistic equation

The logistic difference equation (logistic map) can be written as

$$x_{n+1} = \mu x_n (1 - x_n) = f(x_n) \tag{5.2-1a,b}$$

where x_n is the normalized population with $0 \le x_n \le 1$.

Equation (5.2-1) is a recursive equation and with an initial value x_0, we can generate x_n for all values of n.

Table 5.2-1 presents the values of x_n for various values of μ.

TABLE 5.2-1

Values of x_n for various values of μ

μ $\backslash$ n	0.5	2.0	3.1	3.9	3.9
0	0.250	0.250	0.250	0.250	0.260
1	0.094	0.375	0.581	0.731	0.750
2	0.042	0.469	0.754	0.766	0.731
3	0.020	0.498	0.574	0.698	0.768
4	0.010	0.500	0.758	0.822	0.696
5	0.005		0.569	0.571	0.826
6	0.002		0.760	0.955	0.561
7	0.001		0.565	0.166	0.961
8	0.001		0.762	0.540	0.148
9	0.000		0.562	0.969	0.491
10			0.763	0.119	0.975
11			0.561	0.408	0.096
12			0.763	0.942	0.339
13			0.561	0.214	0.874
14			0.763	0.656	0.428
15			0.561	0.880	0.955

The properties of Equation (5.2-1a) were discussed in Chapter 4. We recall that the equilibrium points x_e are given by

$$x_e = 0, \qquad x_e = 1 - 1/\mu \qquad\qquad (5.2\text{-}2a,b)$$

To determine the stability of the equilibrium points, we need to evaluate df/dx. From Equations (5.2-1a,b), we deduce that

$$\frac{df}{dx} = \mu(1 - 2x_n) \tag{5.2-3}$$

Evaluating at the equilibrium points yields

$$\frac{df}{dx}\bigg|_0 = \mu, \qquad \frac{df}{dx}\bigg|_{1-(1/\mu)} = 2 - \mu \tag{5.2-4a,b}$$

We deduce from the criterion stated in Chapter 4 that if $\mu < 1$, $x_e = 0$ is a stable equilibrium and the other equilibrium point is unstable. Note also that since x_e has to be positive, the equilibrium point $1 - 1/\mu$ does not exist if $\mu < 1$. If $\mu > 1$, $x_e = 0$ is unstable and the point $1 - 1/\mu$ is stable if $1 < \mu < 3$. Thus, if $\mu < 1$, $x_n \longrightarrow 0$ as $n \longrightarrow \infty$ and if $1 < \mu < 3$, $x_n \longrightarrow 1 - 1/\mu$ as $n \longrightarrow \infty$. This is confirmed in Table 5.2-1. If $\mu > 3$, the situation is more complex. From Table 5.2-1, we observe that if $\mu = 3.3$, the values of x_n fluctuate between two values, alternately around 0.76 and 0.56. It seems that there is a periodic solution: in one generation, the population is high and in the next generation, it is low. This variability is confirmed for $\mu = 3.9$, where the values of x_n fluctuate between high, medium, and low. This behavior can be interpreted in terms of bifurcation and will be considered in Section 5.4.

Equation (5.2-1a) can be solved exactly if $\mu = 4$ (Kadanoff, 1983). We rewrite the variable x_n as

$$x_n = (1 - \cos 2\pi\theta_n)/2 \tag{5.2-5}$$

Substituting Equation (5.2-5) into Equation (5.2-1a) yields

$$(1 - \cos 2\pi\theta_{n+1})/2 = 4\left[(1 - \cos 2\pi\theta_n)/2\right]\left[(1 + \cos 2\pi\theta_n)/2\right] \tag{5.2-6a}$$

$$= 1 - \cos^2 2\pi\theta_n \tag{5.2-6b}$$

$$= (1 - \cos 4\pi\theta_n)/2 \tag{5.2-6c}$$

The solution of Equation (5.2-6c) is

$$\theta_{n+1} = 2\theta_n \tag{5.2-7}$$

It follows that

$$\theta_n = 2^n \theta_0 \tag{5.2-8}$$

where θ_0 is the initial value of θ_n.

Since $\cos$ is a periodic function of period 2π, adding an integer to θ_n leads to the same value of x_n. That is to say, we need to consider the fractional part of θ_n only. Equation (5.2-8) suggests that it is more appropriate to express θ_0 in a binary system rather than in a decimal system. At each iteration (multiplication by 2), the "decimal" point is shifted by one place. If θ_0 consists of N digits, precision is lost after N iterations. Alternatively, we can express this loss of precision as follows. Consider two initial values x_0 and x_0' that differ by ε. After N iterations, they generate x_N and x_N' respectively. The difference between x_N and x_N' has grown to $2^N \varepsilon$. For a sufficient large N, x_N can be significantly different from x_N', even if ε is very small.

We observe from the last two columns of Table 5.2-1 that initially the values of x_n are close, but after the third iteration they are significantly different. The fluctuation in the values of x_n is maintained for values of $n > 3$.

This example shows that even a simple non-linear difference equation can give rise to chaos.

Example 5.2-2. **Lorenz's equations**

Lorenz (1963) considered a thin layer of fluid with the lower surface kept at temperature T_0 and the upper surface at temperature T_1 ($< T_0$). Due to buoyancy, the fluid tends to rise and this is opposed by the viscous force. This is the Raleigh-Bénard problem and the dimensionless control parameter is the Raleigh number Ra. If Ra is less than a critical number Ra_c, no convection occurs and if $Ra \geq Ra_c$, Bénard cells appear. These cells are steady cells and are described in Chandrasekhar (1961). If Ra is increased beyond Ra_c, the motion becomes much more complicated and it is in this non-linear region where Lorenz's equations apply.

The derivation of Lorenz's equation is given in the original paper and in Sparrow (1982) and they can be written as

$$\dot{x} = p\,(y - x) = f_1 \qquad\qquad\qquad (5.2\text{-}9a,b)$$

$$\dot{y} = r\,x - x\,z - y = f_2 \qquad\qquad\qquad (5.2\text{-}9c,d)$$

$$\dot{z} = x\,y - b\,z = f_3 \qquad\qquad\qquad (5.2\text{-}9e,f)$$

where the dot denotes differentiation with respect to time (dimensionless), x is associated with the stream function, y is proportional to the temperature difference between the ascending and descending currents at a fixed height, z is proportional to the deviation of the vertical temperature from the linear profile, p is the Prandtl number, $r = Ra / Ra_c$, and b is the ratio of the thickness of the fluid layer to the horizontal size of the convection cell.

The equilibrium points of Equations (5.2-9a to f) are the origin $(0, 0, 0)$ and

$$x = y, \qquad z = r - 1, \qquad x = \pm \sqrt{b\,(r - 1)} \qquad\qquad (5.2\text{-}10a,b,c)$$

From Equation (5.2-10c), we note that if $r < 1$, the origin is the only equilibrium point and it is a stable equilibrium point. This corresponds to the situation when there are no convection cells. The point $r = 1$ is a bifurcation point and for $r > 1$, we have three equilibrium points. To determine the stability of these points, we have to examine the Jacobian matrix $\underline{\underline{J}}$. From Equations (5.2-9a to f), we obtain

$$\underline{\underline{J}} = \text{grad } \underline{f} = \begin{bmatrix} -p & p & 0 \\ r-z & -1 & -x \\ y & x & -b \end{bmatrix} \qquad (5.2\text{-}11\text{a,b})$$

Evaluating $\underline{\underline{J}}$ at the origin yields

$$\underline{\underline{J}}_0 = \begin{bmatrix} -p & p & 0 \\ r & -1 & 0 \\ 0 & 0 & -b \end{bmatrix} \qquad (5.2\text{-}12)$$

The eigenvalues of $\underline{\underline{J}}_0$ are

$$\lambda_1 = -b, \qquad \lambda_{2,3} = -(p+1)/2 \pm [\sqrt{(p+1)^2 + 4p(r-1)}]/2 \qquad (5.2\text{-}13\text{a,b,c})$$

If $r < 1$, all the eigenvalues have negative real parts and as mentioned earlier the origin is stable. If $r > 1$, one of the eigenvalues is positive and the origin is unstable.

Evaluating $\underline{\underline{J}}$ at the two other equilibrium points [Equations (5.2-10a to c)], we find that the eigenvalues are the roots of the equation

$$\lambda^3 + (p+b+1)\lambda^2 + b(p+r)\lambda + 2bp(r-1) = 0 \qquad (5.2\text{-}14)$$

If $r = 1$, the three roots are

$$\lambda_1 = 0, \qquad \lambda_2 = -b, \qquad \lambda_3 = -(p+1) \qquad (5.2\text{-}15\text{a,b,c})$$

Since none of the roots has positive real part, the two equilibrium points are stable and $r = 1$ is a pitchfork bifurcation point. For $r > 1$, we have only one real negative root and two complex conjugate roots with negative real parts. The next bifurcation point is when the real part of the complex root is zero. To find the value of r at which this is possible, we let

$$\lambda = i\lambda_0 \qquad (5.2\text{-}16)$$

Substituting Equation (5.2-16) into Equation (5.2-14) and comparing the real and the imaginary parts, we deduce

$$\lambda_0^2 = 2\,b\,p\,(r-1)\,/\,(p+b+1) \qquad\qquad (5.2\text{-}17\text{a})$$

$$r = r_H = p\,(p+b+3)\,/\,(p-b-1) \qquad\qquad (5.2\text{-}17\text{b})$$

Substituting the value of r_H into Equation (5.2-17a) yields

$$\lambda_0^2 = 2\,b\,p\,(p+1)\,/\,(p-b-1) \qquad\qquad (5.2\text{-}18)$$

Provided $p > b+1$, Equation (5.2-14) can have purely imaginary roots. If $r > r_H$, all three equilibrium points are unstable. It is customary to choose $p = 10$, $b = 8/3$, and from Equation (5.2-17b), we deduce that $r_H \approx 22.74$. In his numerical integration, Lorenz has chosen $r = 28 > r_H$ and $p > b+1$ and we can expect interesting results.

By choosing an appropriate Liapunov function V (see Chapter 3, Problem 6), it is shown that Equations (5.2-9a to f) are stable for all values of r and the trajectories are bounded. Furthermore, div $\underline{f}$ is given by

$$\text{div } \underline{f} = \frac{\partial f_1}{\partial x} + \frac{\partial f_2}{\partial y} + \frac{\partial f_3}{\partial z} = -(p+1+b) \qquad\qquad (5.2\text{-}19\text{a,b})$$

and is negative. This implies that the volume in the phase space is contracting (the system is dissipative) and the trajectories are confined to a bounded region. Since the trajectories cannot intersect, their motion is erratic making a few loops to the right and a few loops to the left (strange attractor). The numerical results obtained by Lorenz confirm this description.

Graphs of computed values of x, y, and z are given in Sparrow (1982). For $r > r_H$, the motion is chaotic and can be shown to be sensitive to the initial conditions. The situation is similar to that in the previous example when $\mu > 3$.

Note that in this example, we have three variables (x, y, z), that is, the phase space is three-dimensional and chaos could happen. In a two-dimensional space, the Poincaré-Bendixson theorem (see Chapter 1) rules out the possibility of chaos.

Experimental support of chaotic motion for the Raleigh-Bénard problem is given in Dubois and Bergé (1981). Due to heavy truncation applied in the derivation of Lorenz's equations, they might not be valid for $r \gg 1$ and there is not necessarily a quantitative agreement between the present analysis and experimental data.

The present example illustrates chaos in a dissipative system (div $\underline{f} < 0$) and in the next example, we consider chaos in a Hamiltonian system.

Example 5.2-3. **Hénon-Heiles model**

This model was introduced by Hénon and Heiles (1964) in their study of the motion of a star inside a galaxy. They used the Hamiltonian formulation. This formulation is the most appropriate one for quantum and statistical mechanics and is widely adopted in the description of chaotic phenomena. In the Hamiltonian description, we introduce a function H, the Hamiltonian of the system. For our purpose, we can regard H as the total (kinetic plus potential) energy of the system. The state of the system is described by the generalized coordinates q_i and the generalized momenta p_i. The Hamiltonian denoted by H is generally a function of all p_i and q_i. If H is not a function of time t, H is a constant. The equations of motion (Hamilton's equations) are written as

$$\dot{q}_i = \frac{\partial H}{\partial p_i}, \qquad \dot{p}_i = -\frac{\partial H}{\partial q_i} \tag{5.2-20a,b}$$

The system of Equations (5.2-20a, b) forms a set of first order equations whereas in the Newtonian description, we have a set of second order differential equations.

If H is not a function of q_1, we deduce from Equation (5.2-20b) that

$$\dot{p}_1 = 0 \quad \Rightarrow \quad p_1 = \text{constant} \tag{5.2-21a,b}$$

The coordinate q_1 is a **cyclic coordinate** and p_1 is a **constant of the motion**. If the number of the constants of the motion is the same as the number of degrees of freedom of the system, the system is **integrable**. It is desirable to find a transformation that can transform all coordinates q_i to cyclic coordinates and preserve the structure of Equations (5.2-20a, b). If such a transformation is possible, we write

$$(q_i, p_i) \longrightarrow (\theta_i, J_i) \tag{5.2-22a}$$

$$H(p_i, q_i) \longrightarrow S(J_i) \tag{5.2-22b}$$

Hamilton's equations become

$$\dot{J}_i = -\frac{\partial S}{\partial \theta_i} = 0 \quad \Rightarrow \quad J_i = \alpha_i \text{ (constants)} \tag{5.2-23a,b,c}$$

$$\dot{\theta}_i = \frac{\partial S}{\partial J_i} = \omega_i (\alpha_1, \alpha_2, \dots , \alpha_n) \tag{5.2-24a,b}$$

Integrating Equation (5.2-24b) yields

$$\theta_i = \omega_i t + \text{constant} \tag{5.2-25}$$

The constants in Equations (5.2-23c, 25) can be evaluated from the initial conditions.

The J_i and θ_i are the **action variables** and **angle variables** respectively. It is not easy to find J_i and θ_i for systems with more than one degree of freedom. The transformation (q_i, p_i) to (θ_i, J_i) is the **canonical transformation.** This transformation is discussed in Golstein (1980). An integrable system does not exhibit chaos.

The Hamiltonian considered by Hénon and Heiles (1964) can be written as

$$H = \frac{1}{2}(p_1^2 + p_2^2 + q_1^2 + q_2^2) + [q_1^2 q_2 - (1/3)q_2^3] \tag{5.2-26}$$

H represents the Hamiltonian of two simple harmonic oscillators coupled by the cubic terms inside square brackets. H may also be considered to be the Hamiltonian for the motion of a particle in a two-dimensional space under the action of a potential ϕ given by

$$\phi = \frac{1}{2}(q_1^2 + q_2^2) + [q_1^2 q_2 - (1/3)q_2^3] \tag{5.2-27}$$

From Equations (5.2-20a, b, 26), we deduce that Hamilton's equations are

$$\dot{q}_1 = p_1, \qquad \dot{q}_2 = p_2 \tag{5.2-28a,b}$$

$$\dot{p}_1 = -(q_1 + 2q_1 q_2), \qquad \dot{p}_2 = -(q_2 + q_1^2 - q_2^2) \tag{5.2-28c,d}$$

Equations (5.2-28a to d) can be written in the form of Equation (1.1-2b) if the following transformations are made

$$q_1 = x_1, \quad q_2 = x_2, \quad p_1 = x_3, \quad p_2 = x_4 \tag{5.2-29a-d}$$

Equations (5.2-28a to d) now become

$$\dot{x}_1 = x_3, \qquad \dot{x}_2 = x_4 \tag{5.2-30a,b}$$

$$\dot{x}_3 = -(x_1 + 2x_1 x_2), \qquad \dot{x}_4 = -(x_2 + x_1^2 - x_2^2) \tag{5.2-30c,d}$$

We note that in this example, div $\underline{f}$ is zero (the system is not dissipative) and the volume in the four-dimensional space $(x_1 - x_4)$ is preserved. If the initial value of H is E, at all subsequent times $H = E$ and the trajectories lie on the three-dimensional surface

$$H(x_1, x_2, x_3, x_4) = E \tag{5.2-31}$$

Combining Equations (5.2-26, 29a-d, 31) yields

$$x_3 = [2E - x_4^2 - x_1^2 - x_2^2 - 2x_1^2 x_2 + (2/3) x_2^3]^{1/2} \tag{5.2-32}$$

If (x_1, x_2, x_4) are known, x_3 can be deduced from Equation (5.2-32) and we need to consider the three-dimensional space (x_1, x_2, x_4). If the motion is bounded in this space, the trajectories will cut a given plane in this space repeatedly. Consider the plane $x_1 = 0$, the trajectory of the first round intersects the plane at s_1, as shown in Figure 5.2-1. On the second round, the trajectory passes through point s_2 on the plane and on the nth round through s_n.

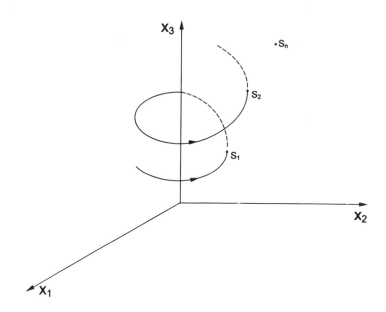

FIGURE 5.2-1   **Poincaré section**

The plane is the **Poincaré section** and was introduced by Poincaré (1892) in his study of the dynamics of the three-body system. The four-dimensional space we started with has been reduced to the two-dimensional Poincaré section. By examining the points s_1, s_2, ... , s_n, we can deduce the properties of the trajectories. To generate the points s_1, s_2, ... , s_n, we start by choosing a value for E and a set of initial conditions $[x_i(0)]$. Note that the chosen initial conditions must satisfy Equation (5.2-31). Equations (5.2-30a-d) are then solved numerically and the points s_1, s_2, ... , s_n are computed. With the same value of E, we choose another set of initial conditions and the corresponding set of s_1, s_2, ... , s_n is computed. If the motion is periodic with only one period, the trajectories coalesce into one closed curve and it intersects the Poincaré section at one point only $(s_1 = s_2 = ... = s_n)$. If the solution is a cycle of period 2, the trajectory intersects the Poincaré section alternately at s_1 and s_2. If the solution has n periods, the trajectory passes through s_1, s_2, ... , s_n in turn and the points s_1, s_2, ... , s_n form a smooth curve. In the present example, it is found that if E is small (say, $E = 1/12$), all the initial conditions lead to smooth closed curves in the

Poincaré section. On increasing the value of E, the structure of the curves formed by s_1, s_2, ... , s_n become more complex and they no longer form smooth curves. At E = (1/6), there is almost no smooth curve and this suggests that the motion is chaotic. If E > (1/6), the motion can be unbounded.

The Hénon-Heiles model has attracted considerable attention and Churchill et al. (1977) have discussed some of the results.

5.3   CRITERIA FOR CHAOS

In this section, we present briefly some methods which can be used to detect the presence of chaos.

Liapunov Exponents

It was stated earlier that chaos can be associated with the exponential divergence of trajectories. We start by considering a one-dimensional system defined by

$$\dot{x} = f(x) \tag{5.3-1}$$

Suppose that we have two trajectories $x_1(t)$ and $x_2(t)$. We denote the distance between them by $s(t)$. The function $s(t)$ can be written as

$$s(t) = x_2 - x_1 \tag{5.3-2}$$

Combining Equations (5.3-1, 2) yields

$$\dot{s} = f(s + x_1) - f(x_1) \tag{5.3-3}$$

Expanding $f(s + x_1)$ and retaining only the linear term in s, Equation (5.3-3) becomes

$$\dot{s} \approx s \left. \frac{df}{dx} \right|_{x_1} \approx \mu s \tag{5.3-4a,b}$$

where μ is a constant and is equal to f' evaluated at $x_1(0)$.

The solution of Equation (5.3-4b) is

$$s(t) = s(0) e^{\mu t} \tag{5.3-5}$$

If $\mu > 0$, $s(t)$ increases exponentially, if $\mu < 0$, $s(t) \longrightarrow 0$ as $t \longrightarrow \infty$, and if $\mu = 0$, $s(t)$ is a constant. Thus μ can be used as a test of the presence of chaos. However, μ is the value of f' evaluated at $x_1(0)$ and Equation (5.3-5) is valid only for t in the neighborhood of $t = 0$. The trajectory $x_1(t)$ is a function of t and f' evaluated at another time t_1 can be different from μ in sign. An average value of f' or a limiting value of f' as $t \longrightarrow \infty$ is preferable. A possible definition of **Liapunov exponent** λ is

$$\lambda = \lim_{t \longrightarrow \infty} \left(\frac{1}{t}\right) \ell n \left[\frac{s(t)}{s(0)}\right] \tag{5.3-6}$$

Note that to determine λ we need to choose a reference trajectory $x_1(t)$ and in general the non-linear differential equation has to be integrated numerically.

For simplicity, we first consider a one-dimensional iterated map which can be written as [Equation (5.2-1b)]

$$x_{n+1} = f(x_n) \tag{5.3-7}$$

Suppose we start the iteration with two initial values x_0 and x_0^* $(= x_0 + \varepsilon)$. After n iterations, x_n and x_n^* differ by

$$s_n = x_n^* - x_n = f^{(n)}(x_0 + \varepsilon) - f^{(n)}(x_0) \tag{5.3-8a,b}$$

Note that $f^{(n)}$ denotes the nth iterate and not the nth derivative.

Equation (5.3-6) can be written in this case as

$$\lambda = \lim_{n \longrightarrow \infty} \left(\frac{1}{n}\right) \ell n \left[\frac{f^{(n)}(x_0 + \varepsilon) - f^{(n)}(x_0)}{\varepsilon}\right] \tag{5.3-9}$$

On letting $\varepsilon \longrightarrow 0$, the terms inside the square bracket represent the derivative of $f^{(n)}$ and Equation (5.3-9) becomes

$$\lambda = \lim_{n \longrightarrow \infty} \left(\frac{1}{n}\right) \ell n \left[\frac{d}{dx} f^{(n)}(x_0)\right] \tag{5.3-10}$$

By definition, $f^{(n)}(x_0)$ can be written as

$$f^{(n)}(x_0) = f(f(f(\dots f(x_0)))) \tag{5.3-11}$$

Taking the derivative and applying the chain rule yields

$$\frac{d}{dx} f^{(n)}(x_0) = f'(x_{n-1}) f'(x_{n-2}) \dots f'(x_0) \tag{5.3-12}$$

where the prime ($'$) denotes the derivative.

Equation (5.3-10) becomes

$$\lambda = \lim_{n \to \infty} \left(\frac{1}{n}\right) \ell n \left[f'(x_{n-1}) \, f'(x_{n-2}) \, ... \, f'(x_0) \right] \tag{5.3-13a}$$

$$= \lim_{n \to \infty} \left(\frac{1}{n}\right) \left[\ell n \, |f'(x_{n-1})| + \ell n \, |f'(x_{n-2})| + ... + \ell n \, |f'(x_0)| \right] \tag{5.3-13b}$$

$$= \lim_{n \to \infty} \left(\frac{1}{n}\right) \left[\sum_{i=0}^{n-1} \ell n \, |f'(x_i)| \right] \tag{5.3-13c}$$

The definition of a Liapunov exponent λ [Equation (5.3-9)] depends on x_0 and different starting values yield different values of λ. We usually compute λ for various values of x_0 and take the average of these values. If the average value of λ is positive, the system is chaotic.

For an n-dimensional discrete system, the recurrence formula is written as

$$\underline{x}_{n+1} = \underline{f}(\underline{x}_n) \tag{5.3-14}$$

where $\underline{x}_n$ and $\underline{f}$ are n-dimensional vectors.

It was shown in Chapter 1, that f' is to be replaced by the Jacobian matrix $\underline{\underline{J}}$ ($= D\underline{f} = \text{grad} \, \underline{f}$) and the properties of the system depend on the eigenvalues of the matrix $\underline{\underline{J}}$. We denote that the Jacobian $\underline{\underline{J}}$ associated with the kth iterate by $\underline{\underline{J}}_k$. We denote the product of $\underline{\underline{J}}_1$, $\underline{\underline{J}}_2$, ..., $\underline{\underline{J}}_p$ by $\underline{\underline{D}}_p$. The n eigenvalues of $\underline{\underline{D}}_p$ are denoted by $\mu_1(p)$, $\mu_2(p)$, ..., $\mu_n(p)$. We define the Liapunov exponents as

$$\lambda_i = \lim_{p \to \infty} \frac{1}{p} \ell n \, |\mu_i(p)|, \qquad i = 1, 2, ..., n \tag{5.3-15}$$

If one of the λ_i values is positive, the system is chaotic. We can proceed in the same way for a n-dimensional continuous system. We recall that the equation governing the dynamical system is given by Equation (1.1-6a) and is reproduced here for convenience

$$\frac{d\underline{x}}{dt} = \underline{f}(\underline{x}) \tag{1.1-6a}$$

We choose two initial conditions $\underline{x}_0$ and $\underline{x}_0^*$ ($= \underline{x}_0 + \underline{\varepsilon}$) and they generate two trajectories $\underline{x}(t)$ and $\underline{x}^*(t)$. The trajectory $\underline{x}(t)$ is the reference trajectory. Let

$$\underline{s}(t) = \underline{x}^* - \underline{x} \tag{5.3-16}$$

Combining Equations (1.1-6a, 5.3-16) and expanding the appropriate expression in a Taylor series yields

$$\dot{\underline{s}} = \underline{s} \cdot \text{grad} \, \underline{f} + O|s|^2 \tag{5.3-17}$$

where grad $\underline{f}$ is evaluated at $\underline{x}(t)$.

The Liapunov exponent λ can be defined by

$$\lambda = \lim_{t \to \infty} \frac{1}{t} \ln \frac{|\underline{s}(t)|}{|\underline{s}(0)|} \tag{5.3-18}$$

Since $\underline{s}(0) \ (= \underline{\varepsilon})$ is a constant, Equation (5.3-18) can be replaced by

$$\lambda = \lim_{t \to \infty} \frac{1}{t} \ln |\underline{s}(t)| \tag{5.3-19}$$

This definition of λ is independent of direction and in a n-dimensional space the trajectories may approach in one direction and diverge in another. Suppose the initial conditions form a n-dimensional sphere and this sphere is deformed to a n-dimensional ellipsoid after a time, because all the components of $\underline{f}$ in Equation (1.1-6a) are not alike. We can define n Liapunov exponents, one for each of the principal directions. The Liapunov exponent λ_i is defined by

$$\lambda_i = \lim_{t \to \infty} \frac{1}{t} \ln a_i(t) \tag{5.3-20}$$

where a_i is the length of the principal axis of the ellipsoid.

The set of λ_i forms the **Liapunov spectrum**.

To determine λ_i, we have to solve Equations (1.1-6a, b, 5.3-16) most likely numerically. If chaos is present, there is a danger of overflow and various precautions have to be taken. Wolf et al. (1985) have described one method of calculating λ_i. This method is quite laborious.

If the system is dissipative (volume contracting), the sum of the Liapunov exponents is negative and this implies that at least one of the λ_i is negative. If one of the λ_i is positive, it implies that the trajectory diverges in this direction and the system is chaotic. In a three-dimensional space, we have three λ_i $(\lambda_1, \lambda_2, \lambda_3)$ and we order them as $\lambda_1 \geq \lambda_2 \geq \lambda_3$. If all three λ_i are negative, we have a fixed attractor; if $\lambda_1 = 0$, $\lambda_3 < \lambda_2 < 0$, we have a limit cycle; and if $\lambda_1 = \lambda_2 = 0$, $\lambda_3 < 0$, we have a torus.

Argyris et al. (1994) have calculated λ_1 for the Lorenz system (Example 5.2-2) for various values of r. In the range $1 < r < 24$, λ_1 is negative and the system is non-chaotic (a fixed attractor). If $24 \leq r \leq 146.5$, λ_1 is positive except for a few short intervals where λ_1 is zero. In this case, we have a chaotic system with occasional windows of periodic solutions. A wide window of periodic solution is present in the interval $146.5 \leq r \leq 166$, and for $r > 166$, chaotic motion dominates again.

The Liapunov exponents can also be calculated from experimental data and this is described in Wolf et al. (1985). The presence of chaos can also be determined by the method of power spectrum and this is described in the next section.

Power Spectrum and Autocorrelation

We first review some of the properties of Fourier series. The Fourier series of a function $x(t)$ can be written as

$$x(t) = \sum_{n=-\infty}^{\infty} c_n e^{int} \tag{5.3-21a}$$

$$c_n = \frac{1}{2\pi} \int_{-\infty}^{\infty} x(t) e^{-int} dt \tag{5.3-21b}$$

Here, we have assumed that the period is 2π and if the period tends to infinity, we replace the Fourier series by the Fourier integral.

The function $x(t)$ is represented by

$$x(t) = \int_{-\infty}^{\infty} A(\omega) e^{i\omega t} dt \tag{5.3-22a}$$

$$A(\omega) = \frac{1}{2\pi} \int_{-\infty}^{\infty} x(t) e^{-i\omega t} dt \tag{5.3-22b}$$

The quantities c_n and $A(\omega)$ are complex and are the Fourier transforms of $x(t)$.

The time average of the square of $x(t)$ is given by

$$\langle x^2 \rangle = \frac{1}{2\pi} \int_{-\pi}^{\pi} |x(t)|^2 dt = \sum_{n=-\infty}^{\infty} |c_n|^2 \tag{5.3-23a,b}$$

The quantity c_n^2 is the **power spectrum** associated with the harmonic number n. In the case of the Fourier integral, the power spectrum is $|A(\omega)|^2$.

The **correlation** of two functions $x_1(t)$ and $x_2(t)$ is defined by

$$\text{corr } (x_1, x_2) = \int_{-\infty}^{\infty} x_1 (t + \tau)\, x_2(t)\, dt \tag{5.3-24}$$

The **autocorrelation** is the correlation of a function with itself and is written as

$$\text{corr } (x, x) = a\,(\tau) = \int_{-\infty}^{\infty} x\,(t + \tau)\, x\,(t)\, dt \tag{5.3-25a,b}$$

The **mean autocorrelation** $\langle a\,(\tau) \rangle$ is more frequently employed than $a\,(\tau)$. It is defined by

$$\langle a\,(\tau) \rangle = \lim_{T \to \infty} \frac{1}{2T} \int_{-T}^{T} x\,(t + \tau)\, x\,(t)\, dt \tag{5.3-26}$$

If the system has a fixed point attractor, $x\,(t)$ is a constant $(= x_e)$. From Equation (5.3-26), we deduce that

$$\langle a\,(\tau) \rangle = x_e^2 \tag{5.3-27}$$

For a limit cycle (periodic solution, period $2\pi/\omega$), $x\,(t)$ can be represented by its Fourier series, that is

$$x\,(t) = \sum_{n=-\infty}^{\infty} c_n\, e^{i n \omega t} \tag{5.3-28}$$

Substituting Equation (5.3-28) into Equation (5.3-26) yields

$$\langle a\,(\tau) \rangle = \sum_{n=-\infty}^{\infty} |c_n|^2 \cos n\omega\tau \tag{5.3-29}$$

White noise is completely random and there is no relationship between the values $x\,(t)$ at various times. In this case, $\langle a\,(\tau) \rangle$ is zero everywhere except at $\tau = 0$, that is, it is represented by the Dirac delta function δ.

In the case of chaotic behavior, $\langle a\,(\tau) \rangle$ is initially non-zero, there is a correlation between two close values of $x\,(t)$, and as τ increases, $\langle a\,(\tau) \rangle$ diminishes to zero in an irregular way. The power spectrum consists of a wide continuous spread of frequencies. Table 5.3-1 gives the values of $\langle a\,(\tau) \rangle$ and of the power spectrum $P\,(\omega)$ $(|A\,(\omega)|^2$ or $|c_n|^2)$ for various cases.

Further details on power spectrum and autocorrelation are given in Farmer et al. (1980) and Kadanoff (1983).

TABLE 5.3-1

Average autocorrelation $\langle a(\tau)\rangle$ and power spectrum $P(\omega)$ for various cases

$x(t)$	$\langle a(\tau)\rangle$	$P(\omega)$
Fixed point attractor $\quad x_e$	x_e^2	$2\pi\, x_e^2\, \delta(\omega)$
Limit cycle $\quad \displaystyle\sum_{n=-\infty}^{\infty} c_n\, e^{in\omega t}$	$\displaystyle\sum_{n=-\infty}^{\infty} |c_n|^2 \cos n\omega\tau$	$2\pi \displaystyle\sum_{n=-\infty}^{\infty} |c_n|^2\, \delta(\omega - n\omega)$
Torus $\quad \displaystyle\sum_{\substack{n=-\infty \\ m=-\infty}}^{\infty\;\infty} c_{nm}\, e^{i\alpha_{nm}t}$ $\alpha_{nm} = n\omega_1 + m\omega_2$	$\displaystyle\sum_{\substack{n=-\infty \\ m=-\infty}}^{\infty\;\infty} |c_{nm}|^2 \cos \alpha_{nm} t$	$2\pi \displaystyle\sum_{\substack{n=-\infty \\ m=-\infty}}^{\infty\;\infty} |c_{nm}|^2\, \delta(\omega - \alpha_{nm})$
White noise	$a_0\, \delta(\tau)$ a_0 is a constant	P_0 is a constant

Fractal Dimension

In Chapter 1, we have described the possibility of a strange attractor in a three-dimensional space. The space occupied by the trajectories is greater than two and is not quite the full three-dimensional space. We can now introduce a fractional dimension and this concept of **fractal** was developed by Mandelbrot (1982). One of the properties of fractals is **self-similarity**, that is, the recurrence of the same pattern over a range of scales. An example of a fractal is the **Cantor set** which is shown in Figure 5.3-1. This set is constructed in the following way. We start with a line of unit length and remove the middle third. The middle third of each of the remaining ends is again deleted and this process in continued ad infinitum. Ultimately, we obtain an infinite, non-countable set of limit points. Note the self-similarity. The original line is one-dimensional. What is the dimension of the remaining line segments after several deletions? The answer depends on the definition of dimension.

Several definitions of dimensionality are currently in use and a discussion of the various definitions can be found in Farmer et al. (1983). Here, we consider only the **capacity dimension** and the **correlation dimension**.

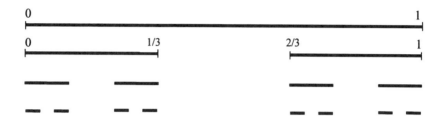

FIGURE 5.3-1   Cantor set

The capacity dimension is also known as the **box-counting dimension** because it involves counting the number of boxes required to fill the space whose dimension we wish to determine. Suppose we are required to determine the dimension of a one-dimensional object. We use a ruler (line segment) of length ε and count the number of $N(\varepsilon)$ of rulers needed to contain the whole object. Similarly, if the object is two or three-dimensional, we use a square of size ε^2 or a cube of size ε^3 and count the number of boxes (line segments, squares, or cubes) required to cover the whole object. This process can be extended to n-dimensional objects. The **capacity dimension** D_c is defined by

$$D_c = - \lim_{\varepsilon \to 0} \frac{\ln N(\varepsilon)}{\ln \varepsilon} \qquad (5.3\text{-}30)$$

We calculate D_c for a unit square. The number of squares $N(\varepsilon)$ of size ε^2 required to cover the unit square is $1/\varepsilon^2$. Equation (5.3-30) becomes

$$D_c = - \lim_{\varepsilon \to 0} \frac{\ln (1/\varepsilon^2)}{\ln \varepsilon} = \lim_{\varepsilon \to 0} \frac{2\ln \varepsilon}{\ln \varepsilon} = 2 \qquad (5.3\text{-}31\text{a,b,c})$$

The capacity dimension of a square is two, as expected.

We now determine D_c for the Cantor set. At the nth stage of the subdivision of the unit line, we have 2^n segments each of length $(1/3)^n$. We choose $\varepsilon = (1/3)^n$ and the number $N(\varepsilon)$ of boxes of size ε needed to cover the Cantor set is 2^n. From Equation (5.3-30), we deduce that

$$D_c = -\lim_{n \to \infty} \frac{\ln 2^n}{\ln (1/3)^n} = \lim_{n \to \infty} \frac{n \ln 2}{n \ln 3} = \frac{\ln 2}{\ln 3} = 0.6309 \qquad (5.3\text{-}32\text{a-d})$$

The capacity dimension of the Cantor set is a fraction.

To determine the capacity dimension of an attractor, we partition the phase space into boxes, count the number of boxes, and take the limit as indicated in Equation (5.3-30). If the dimension is a non-integer, the attractor is strange. In most cases, the trajectories are obtained numerically or are experimental data, they are not exact and taking the limit as $\varepsilon \to 0$ can be problematic. For practical calculations, the correlation dimension is more appropriate.

For simplicity, we introduce the correlation dimension D_k for the one-dimensional case and then generalize to n dimensions. Consider N points on a trajectory which has evolved for a long time so that all transients have decayed. We focus on a point i (x_i) and count the number of points on the trajectory that are within a distance ε from x_i. We denote this number by $N_i(\varepsilon)$ and the point x_i is not included in the calculation of N_i. The relative number of points $p_i(\varepsilon)$ that are within the distance ε from x_i is

$$p_i(\varepsilon) = N_i / (N - 1) = (1/N - 1) \sum_{j=i}^{N} H(\varepsilon - |x_i - x_j|) \qquad (5.3\text{-}33\text{a,b})$$

where H is the Heaviside step function.

We recall that $H(\varepsilon - |x_i - x_j|)$ is one if $\varepsilon > |x_i - x_j|$ and is zero if $\varepsilon < |x_i - x_j|$. It follows that Equations (5.3-33a, b) are equivalent.

The **correlation sum** $C(\varepsilon)$ is defined by

$$C(\varepsilon) = \frac{1}{N} \sum_{i=1}^{N} p_i(\varepsilon) \qquad (5.3\text{-}34)$$

The **correlation dimension** D_k is

$$D_k = \lim_{\varepsilon \to 0} \frac{\ln C(\varepsilon)}{\ln \varepsilon} \qquad (5.3\text{-}35)$$

In extending to the n-dimensional case, x_i is replaced by the vector $\underline{x}_i$ and we have to define the distance $|x_i - x_j|$. We may choose the Euclidean distance or the sum of the absolute values of the differences of the individual components of $|x_i - x_j|$. The number $N_i(\varepsilon)$ is now the number of points that are within a sphere of radius ε with $\underline{x}_i$ as a center.

Grassberger and Procaccia (1983) have shown that

$$D_k \leq D_c \tag{5.3-36}$$

That is, D_k is the lower limit of D_c. If D_k is not an integer, the attractor is strange.

Argyris et al. (1994) have calculated D_k for the Lorenz system (Example 5.2-2).

Poincaré Section

In Example 5.2-3, we have introduced the Poincaré section so as to facilitate the understanding of the dynamics of the Hénon-Heiles system. We recall that a Poincaré section is a section that cuts transversally through a bundle of trajectories, as shown in Figure 5.2-1. A trajectory cuts through a Poincaré section repeatedly and we denote the successive points of intersection by s_1, s_2, ... , s_n. Instead of studying the entire trajectory, we need to consider only the points s_1, s_2, ... , s_n. There is a relationship between the successive points and we can express this by writing

$$s_{n+1} = f(s_n) \tag{5.3-37}$$

where the **Poincaré (return) map** f depends on the equations describing the system and the choice of the Poincaré section.

The introduction of a Poincaré section replaces the original differential equation by a difference equation [Equations (1.1-2a) by (5.3-37)], reduces the dimension of the space (in Example 5.2-3, from three to two). Experimental data can be analyzed via a Poincaré section and in this case, it leads to a reduction in the size of data because we need only to consider the data near and on the Poincaré's section. However, it is as difficult to deduce an exact analytical form for f as it is to solve the original differential equation analytically. In most cases, the points s_1, s_2, ... , s_n are obtained numerically or are experimental data. Hénon (1982) has suggested a numerical scheme for solving the differential equation and for generating the Poincaré section simultaneously.

The pattern of the points s_1, s_2, ... , s_n depends on the attractor. It has already been mentioned in Example 5.2-3 that for a periodic solution with only one period, the Poincaré section consists of only one point. If the phase space is two-dimensional, the Poincaré section is a line. The trajectory of a periodic solution is a closed curved and it intersects the Poincaré section (a line) repeatedly at the same point. If the solution has n periods, the Poincaré section has n points, each point corresponding to one period.

In a three-dimensional phase space, we can have a type of motion that does not exist in lower dimensional space. This is the quasi-periodic motion. It can be represented by two oscillators vibrating at two different frequencies. The trajectories are confined to the surface of a torus (Example 5.2-3). The equation of a torus, as shown in Figure 5.3-2, can be written as

$$x_1 = (R + r \sin \omega_1 t) \cos \omega_2 t \tag{5.3-38a}$$

$$x_2 = r \cos \omega_1 t \tag{5.3-38b}$$

$$x_3 = (R + r \sin \omega_1 t) \sin \omega_2 t \tag{5.3-38c}$$

where ω_1 and ω_2 are the two frequencies, r and R are as shown in Figure 5.3-2.

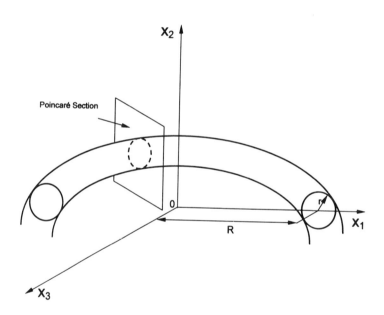

FIGURE 5.3-2   Surface of a torus and a Poincaré section

The type of motion depends on the ratio ω_1 / ω_2 $(= w)$ and w is the **winding number**. If w is a rational number, the trajectories are closed curves and the Poincaré section consists of a finite number of points. If w is an irrational number, the trajectories are not closed, they fill the whole surface of the torus and they generate a continuous curve in the Poincaré section. This is the **quasi-periodic** motion, a motion that never exactly repeats itself and is not chaotic. Similarly, in higher dimensional space $(n > 3)$ if the frequencies ω_1, ω_2, ... , ω_n do not satisfy the equation

$$n_1 \omega_1 + ... + n_n \omega_n = 0 \tag{5.3-39}$$

for some integers n_i, the trajectories cover the whole surface of the $(n-1)$-dimensional torus. The Poincaré section consists of a closed curve.

If the motion is chaotic, the Poincaré section has a complex geometry with the possibility of fractal dimension. The points s_1, s_2, ... , s_n are located in a haphazard way and they do not form a closed curve.

We can also analyze the motion by considering the Poincaré map f defined in Equation (5.3-37). Note that if the Poincaré section is two-dimensional, the point s_n has coordinates (x_n, y_n) and Equation (5.3-37) is written as

$$x_{n+1} = f_1(x_n, y_n) \tag{5.3-40a}$$

$$y_{n+1} = f_2(x_n, y_n) \tag{5.3-40b}$$

Equation (5.3-37) is a n-dimensional recurrent equation [Equation (5.3-14)]. In a three-dimensional highly dissipative system, due to volume contraction, the two-dimensional Poincaré section reduces to a one-dimensional curve. In this case, we can plot x_{n+1} versus x_n to determine f_1. If the slope of f_1 is everywhere greater than one, the motion is chaotic. If the plotted points (x_{n+1}, x_n) scatter randomly, it also implies chaos and/or the presence of noise.

In Lorenz's model (Example 5.2-2), we may choose the Poincaré section to be

$$xy - bz = 0 \tag{5.3-41}$$

From Equation (5.2-9e), we deduce that the points of intersection s_1, s_2, ... , s_n correspond to the successive maxima of z which we denote by Z_1, Z_2, ... , Z_n. Plotting Z_{n+1} versus Z_n determines f and Lorenz found that for the value of $r = 28$, the slope of f is everywhere greater than one, implying chaotic motion.

The plot of x_{n+1} versus x_n is the **first-order return map**. Similarly, we can plot x_{n+2} versus x_n and obtain the **second-order return map**. Higher order return maps can similarly be generated. Hénon (1976) has suggested that instead of solving the differential equation and deducing the Poincaré map, it is simpler to consider a Poincaré return map directly. By an appropriate choice of parameters, it is possible to show that the differential equation and the map exhibit the same features and that they qualitatively describe the system. Inspired by the numerical calculations on the Lorenz system (Example 5.2-2), Hénon (1976) proposed that the differential equation [Equations (5.2-9a-f) can be simulated by the following map

$$x_{n+1} = 1 + y_n - a x_n^2 \tag{5.3-42a}$$

$$y_{n+1} = b x_n \tag{5.3-42b}$$

where a and b are constants.

This map can be considered to be a sequence of three maps

(i) $x' = x$, $y' = 1 + y - ax^2$ (5.3-43a,b)

(ii) $x'' = bx'$, $y'' = y'$ (5.3-44a,b)

(iii) $x''' = y''$, $y''' = x''$ (5.3-45a,b)

The map given by Equations (5.3-42a, b) is the product of the three mappings defined by Equations (5.3-43a-45b). The effects of the three mappings on an initial ellipse are shown in Figures 5.3-3a, b.

The Jacobian J of the mapping [Equations (5.3-42a, b)] is

$$J = \begin{vmatrix} 1 - 2a & 1 \\ b & 0 \end{vmatrix} = -b \qquad (5.3\text{-}46a,b)$$

This mapping preserves the constant negative divergence of the Lorenz system. The equilibrium points (x_e, y_e) are obtained by solving

$$x_e = 1 + y_e - ax_e^2, \qquad y_e = bx_e \qquad (5.3\text{-}47a,b)$$

The solution is

$$x_e = [-(1-b) \pm \sqrt{(1-b)^2 + 4a}]/2a, \qquad y_e = bx_e \qquad (5.3\text{-}48a,b)$$

The equilibrium points are real if

$$a > -(1-b)^2/4 \qquad (5.3\text{-}49)$$

The two equilibrium points are unstable if

$$a > 3(1-b)^2/4 \qquad (5.3\text{-}50)$$

The term b determines the rate of dissipation and it is usually chosen to be 0.3. For this chosen value of b, the value of a is chosen to be greater than $3(1-b)^2/4$. By trial and error, it has been found that for $a = 1.4$, the mapping simulates the strange attractor of the Lorenz system.

Equations (5.3-42a, b) can be considered to be the most general quadratic mapping with a constant Jacobian (Hénon, 1976).

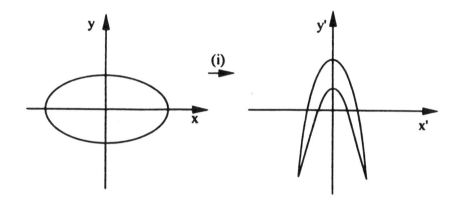

**FIGURE 5.3-3a   Effects of the mapping defined by
Equations (5.3-43a, b) on an ellipse**

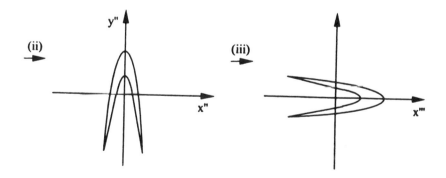

**FIGURE 5.3-3b   Effects of the mappings defined by
Equations (5.3-44a-45b) on an ellipse**

Time-Delay Plot

The state of a dynamical system is described by the trajectories in the phase space and in some cases, the dimension of the space can be so large that it is not possible to generate the trajectories. For example, in a chemical reaction, we may have n species (the phase space is n-dimensional). If n is not small, it is extremely time consuming to measure the concentration of the n species as a function of time. Takens (1981) has shown that the trajectories in the n-dimensional space can be replaced by another curve which can be generated from the time series measurements of a single species which we denote by B. From the time series measurements of $B(t)$, we can plot $B(t_i)$ versus $B(t_i + T)$, where T is a constant time delay, for various values of t_i. The resulting curve is a two-dimensional **time-delay plot**. A m-dimensional time-delay plot of $B(t_i)$, $B(t_i + T)$, ... , $B[t_i + (m-1)T]$ can be drawn with $B(t_i)$, $B(t_i + T)$, ... as coordinate axes; this space is the **embedding space** and m is the **embedding dimension**. If m is chosen large enough, the main dynamical features of the system, such as the dimension of the attractor, the positive Liapunov exponents, calculated from the phase and embedding spaces are the same. A sufficient condition to achieve this is $m = 2n + 1$. In practice, it is desirable to keep m as low as possible.

To determine the minimum embedding dimension, we calculate the correlation dimension D_k [Equation (5.3-35)] for increasing values of m until the values of D_k become constant. The calculations of D_k for the x, y, z components of the Lorenz equations (Example 5.2-2) are given in Argyris (1994). Other methods of finding the minimum embedding dimension have been proposed by (Aleksic, 1991; Kennel et al., 1992). The choice of the time-delay T is in theory arbitrary, but in reality it is not. If T is too small, the change observed can be dominated by noise and if T is too large, there is no relationship between two consecutive data points. For a discrete system, it is usual to take T to be unity. For a continuous system or experimental data points, we can calculate the autocorrelation [Equations (5.3-25a, b)] at successive lags and the optimal lag corresponds to a low value of the autocorrelation. Liebert and Schuster (1989) have proposed other methods of determining T.

The time-delay method was used to analyze the data of a Belousov-Zhabotinskii (BZ) reaction (Roux et al., 1983). The reaction takes place in a continuously stirred tank reactor (CSTR) involving more than thirty chemical constituents. The feed chemicals are malonic acid, potassium bromate, cerous sulphate, and sulfuric acid and the recorded output is the concentration of the bromide ions. The control parameter is the flow rate which is equivalent to the residence time. The lower the flow rate, the longer the residence time.

If the residence time is very short, there is little time for reactions to take place and the chemicals leave the reactor at almost the same feed concentration. If the residence time is very long, chemical equilibrium state can be reached. There is a range of residence times where oscillations and chaos occur and this range of residence times depend on the temperature and the feed concentration. Figure 5.3-4 shows the bromide ion potential time series for (a) periodic, (b) chaotic states observed in the BZ reaction.

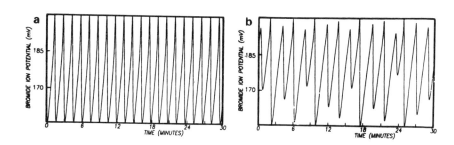

FIGURE 5.3-4   Bromide ion potential B (t) time series observed in the BZ reaction.
(a) periodic state, (b) chaotic state. Adapted from Roux et al. (1983)

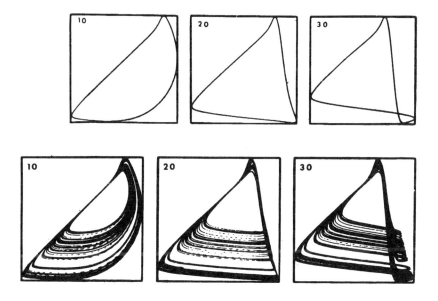

FIGURE 5.3-5   Two-dimensional phase plots, $B(t_i)$ versus $B(t_i + T)$, constructed from
Figure 5.3-4. The values of T are expressed as multiples of the time between successive
measurement (0.88s) and are indicated on each plot. (a) periodic; (b) chaotic states.
Adapted from Roux et al. (1983)

Figure 5.3-5 shows the constructed curves of $B(t_i + T)$ versus $B(t_i)$, where B is the bromide ion potential computed from the data of Figure 5.3-4 for three values of T. Note that in the case of periodic state, we obtain a single closed curve. In the chaotic case, no closed curve is obtained. Even at such low embedding dimension, it is possible to detect the possibility of chaos.

To be convinced of the presence of a chaotic state, Roux et al. (1983) plotted a three-dimensional trajectory, $B(t_i)$, $B(t_i + T)$, and $B(t_i + 2T)$, took a Poincaré section, obtained the Poincaré return map, and calculated the Liapunov exponent λ_i. They found that the value of λ_i is 0.3 ± 0.1 which indicates chaotic behavior.

Several mathematical models, under the name of **Oregonator**, of the BZ reaction have been proposed. The numerical simulations yield a periodic-chaotic sequence confirming the experimental data (Argoul et al., 1987).

In the next section, we discuss various routes to chaos.

5.4   ROUTES TO CHAOS

In the previous section, we have considered several methods that have been used to identify a chaotic motion. In the present section, we describe the possible routes that a regular motion may take on its way to chaos.

Period Doubling

To illustrate this type of transition, we reconsider the logistic equation [Equations (5.2-1a, b)]. In Example 5.2-1, we have observed that if $0 < \mu < 1$, the origin ($x = 0$) is a stable point. If $1 < \mu < 3$, the equilibrium point $x = 1 - 1/\mu$ is stable and the origin is unstable. If $\mu > 3$, the point $x = 1 - 1/\mu$ is unstable. From Table 5.2-1, we note that for $\mu = 3.1$, x converges to two values ($x_1^* \approx 0.76$ and $x_2^* \approx 0.56$) alternately. The points x_1^* and x_2^* are two attractors and the system has two periods. At $\mu = \mu_1$ ($= 3$), the solution bifurcates into two periodic solutions (x_1^* and x_2^*) as shown in Figure 5.4-1.

The solution x_1^* generates x_2^* and x_2^* generates x_1^* and this can be expressed mathematically by

$$x_1^* = f(x_2^*) = \mu x_2^* (1 - x_2^*) \qquad \text{(5.4-1a,b)}$$

$$x_2^* = f(x_1^*) = \mu x_1^* (1 - x_1^*) \qquad \text{(5.4-1c,d)}$$

Combining Equations (5.4-1a, c) yields

$$x_1^* = f^{(2)}(x_1^*), \qquad x_2^* = f^{(2)}(x_2^*) \qquad \text{(5.4-2a,b)}$$

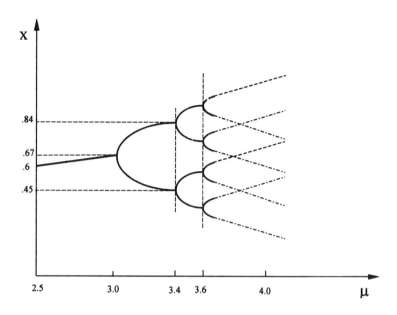

FIGURE 5.4-1   Bifurcation for the logistic equation

The points x_1^* and x_2^* are the equilibrium points of $f^{(2)}$ (second iterate). We denote the equilibrium points of $f^{(2)}$ by x_e^* and they are given by

$$x_e^* = f^{(2)}(x_e^*) \tag{5.4-3}$$

From Equations (5.4-1b, d), we deduce that

$$f^{(2)}(x_e^*) = \mu^2 x_e^*(1 - x_e^*)[1 - \mu x_e^*(1 - x_e^*)] \tag{5.4-4}$$

Substituting Equation (5.4-4) into Equation (5.4-3) yields

$$x_e^* \left\{ (\mu x_e^* - \mu + 1)[\mu^2 x_e^{*2} - (\mu^2 + \mu) x_e^* + (1 + \mu)] \right\} = 0 \tag{5.4-5}$$

The four roots are

$$0, \quad 1 - 1/\mu, \quad [1 + \mu + \sqrt{\mu^2 - 2\mu - 3}]/2\mu, \quad [1 + \mu - \sqrt{\mu^2 - 2\mu - 3}]/2\mu$$

The first two roots $(0, 1 - 1/\mu)$ were obtained earlier and we denote the next two roots by x_1^* and x_2^*. On setting $\mu = 3.1$, we obtain $x_1^* \approx 0.76$ and $x_2^* \approx 0.56$ in agreement with the values given in Table 5.2-1.

To determine the stability of the equilibrium points, we have to evaluate $f^{(2)'}$ which is given by Equation (5.3-12). For the points x_1^* and x_2^*

$$f^{(2)'} = \frac{d}{dx} f^{(2)} = f'(x_1^*) f'(x_2^*) \tag{5.4-6a,b}$$

$$= \mu(1 - 2x_1^*)(1 - 2x_2^*) \tag{5.4-6c}$$

$$= -(\mu^2 - 2\mu - 4) \tag{5.4-6d}$$

We deduce that $|f^{(2)'}| > 1$ if $\mu > 1 + \sqrt{6}$, that is, if $\mu > 1 + \sqrt{6}$ (≈ 3.45), x_1^* and x_2^* are unstable. At $\mu = \mu_2$ (≈ 3.45), both x_1^* and x_2^* bifurcate, each in two other solutions and the system has four periodic solutions (or is of period four). The equilibrium points x_e^{**} are obtained by solving

$$x_e^{**} = f^{(4)}(x_e^{**}) \tag{5.4-7}$$

Equation (5.4-7) has six roots, four roots are known $(0, 1 - 1/\mu, x_1^*, x_2^*)$ and we denote the two additional roots by x_3^* and x_4^*. For $\mu > \mu_2$, the solution oscillates between x_1^*, x_2^*, x_3^* and x_4^* (period four). As μ is further increased, x_1^* to x_4^* become unstable and proceeding in this manner, at $\mu = \mu_3$ (≈ 3.54), each of x_1^* to x_4^* bifurcates into two periodic solutions resulting in a system of period eight. This system becomes unstable at $\mu = \mu_4$ (≈ 3.56) producing a system of period sixteen. This process of instability and period-doubling continues as μ is increased and the interval between successive μ_i at which period-doubling occurs decreases rapidly. The μ_i tend to a finite limit μ_∞ (≈ 3.5699) and beyond that value the motion is chaotic. The region $3.57 \le \mu \le 4$ is characterized by irregular, chaotic behavior with occasional windows of periodic motion, one of which is at $\mu = 1 + \sqrt{8}$.

The ratio δ defined by

$$\delta = \lim_{n \to \infty} \left(\frac{\mu_n - \mu_{n-1}}{\mu_n + 1 - \mu_n} \right) \approx 4.669201 \tag{5.4-8a,b}$$

is the **Feigenbaum constant**.

The Feigenbaum constant is a universal constant and is valid for all quadratic maps with a maximum (Feigenbaum, 1978) and is not confined to logistic maps only. It is a measure of the rate of convergence of μ_i. Hilborn (1994) has discussed the origins and the importance of universal constants associated with Feigenbaum. The transition to chaos via period-doubling has been observed experimentally. Libchaber et al. (1983) have studied the routes to chaos in a Rayleigh-Bénard experiment (Example 5.2-2) where mercury is confined between two parallel plates and is heated from below. In addition, a horizontal magnetic field B_0 is applied to the fluid. The two control parameters are the Rayleigh number Ra $(= g\alpha d^3 \Delta T / k\nu)$ and the Chandrasekhar number Q $(= \sigma B_0^2 d^2 / \rho\nu)$. Here, g is gravity, α is the coefficient of thermal expansion, d is the spacing between the plates, ΔT is the temperature difference between the upper and the lower plates, k is the thermal diffusivity, ν is the kinematic viscosity, σ is the electrical conductivity, and ρ is the density. The observable quantity is the temperature. The experimental procedure involves increasing Ra slowly at constant Q and the temperature profile at the upper plate is observed. Convection (first instability) starts at a critical Rayleigh number Ra_c. Bifurcations to periods 4, 8, and 16 were observed at $Ra/Ra_c = 3.52$, 3.62, and 3.65 respectively, for Q = 22. The computed Feigenbaum number δ was found to be 4.4 ± 0.1. Results for other values of Q are also reported.

The period-doubling route to chaos has also been observed in other experimental settings such as in non-linear electrical oscillators (Testa et al., 1982), optical systems (Gibbs et al., 1981), and in chemical reactions (Coffman et al., 1987).

Intermittency

The **intermittency** route to chaos is characterized by a periodic motion with occasional bursts of irregular behavior. On increasing the values of the control parameter, these occasional bursts occur more frequently and eventually the motion is chaotic. This is similar to the transition from laminar to turbulent flow in a pipe. However, in fluid mechanics, the occurrence of the transition has a spatio-temporal character. Here, we are concerned with temporal processes only.

In the previous section (period-doubling), we have mentioned that the logistic map has a window of periodic motion in the neighborhood of $\mu = 1 + \sqrt{8}$. Argyris et al. (1994) have shown that the route to chaos in this case can be via intermittency.

Intermittent behavior is usually associated with the loss of stability of a periodic motion. The existence of a periodic motion can be ascertained from the Poincaré section. For a three-dimensional motion, the Poincaré map is given by Equations (5.3-40a, b). The existence of a periodic solution implies that the mapping has fixed points (x_e, y_e) and these points satisfy

$$x_e = f_1(x_e, y_e), \qquad y_e = f_2(x_e, y_e) \qquad\qquad (5.4\text{-}9a,b)$$

The stability of the points (x_e, y_e) depends on the eigenvalues of the Jacobian matrix $\underline{\underline{J}}$ defined by

$$\underline{\underline{J}} = \begin{bmatrix} \dfrac{\partial f_1}{\partial x} & \dfrac{\partial f_1}{\partial y} \\[2ex] \dfrac{\partial f_2}{\partial x} & \dfrac{\partial f_2}{\partial y} \end{bmatrix} \qquad\qquad (5.4\text{-}10)$$

The derivatives are evaluated at (x_e, y_e) and $\underline{\underline{J}}$ is the **Floquet matrix**. The eigenvalues of $\underline{\underline{J}}$ (λ_1, λ_2) are the **Floquet multipliers**. The points (x_e, y_e) are stable if $|\lambda_i| < 1$ $(i = 1, 2)$, that is, the periodic solutions are stable. The periodic solution loses its stability if the absolute value of one of the λ_i is greater than one.

Three types of intermittency, depending on λ_i, have been described. In types I and III, λ_1 and λ_2 are real and in type II, λ_1 and λ_2 are complex conjugates. If λ_1 (or λ_2) crosses the unit circle at 1, it is of type I. If the crossing occurs at -1, the behavior is of type III. The bifurcation at λ (λ_1 or λ_2) $= 1$ can be a saddle-node, a transcritical, or a pitchfork and these bifurcations are defined in Chapter IV. Complex eigenvalues correspond to a Hopf bifurcation.

Systems that exhibit type I intermittency may also exhibit period-doubling route to chaos. Type I intermittency has been observed in a Rayleigh-Bénard convection experiment (Bergé et al., 1980) and in chemical reactions (Pomeau et al., 1981).

To illustrate type I intermittency, we consider the map

$$x_{n+1} = f(x_n) = \mu + x_n + x_n^2 \tag{5.4-11}$$

If μ is negative, the two equilibrium points (x_{1e}, x_{2e}) are

$$x_{1e} = \sqrt{-\mu}, \qquad x_{2e} = -\sqrt{-\mu} \tag{5.4-12a,b}$$

To determine the stability of x_{1e} and x_{2e}, we compute f'. From Equations (5.4-11, 12a, b), we deduce that

$$f'(x_{1e}) = 1 + 2\sqrt{-\mu}, \qquad f'(x_{2e}) = 1 - 2\sqrt{-\mu} \tag{5.4-13a, b}$$

It follows that x_{1e} is unstable and x_{2e} is stable. If μ is negative, the iterates converge towards x_{2e}. If $\mu = 0$, x_{1e} and x_{2e} coalesce to the origin $(x_{1e} = x_{2e} = 0)$ and this point is semi-stable. If the initial value x_0 is negative, the iterates x_n will tend to the origin and if $x_0 > 0$, x_n will diverge from the origin. If $\mu > 0$, x_{1e} and x_{2e} vanish, Equation (5.4-11) has no equilibrium point and the mapping function does not intersect the curve (bisector) $x_{n+1} = x_n$. Near the bifurcation point (origin), there is a narrow tunnel between the bisector and the mapping function as shown in Figure 5.4-2.

As the iterates move through the narrow tunnel, the system remains near a fictional fixed point and the motion is periodic. Once the iterates have moved out of the tunnel, the motion is chaotic. After a lapse of time, the trajectory may be recaptured into the tunnel resulting in another time interval of periodic motion. We have an interval of periodic followed by chaotic motions and the distribution of chaotic or periodic motion depend on the global structure of the phase space.

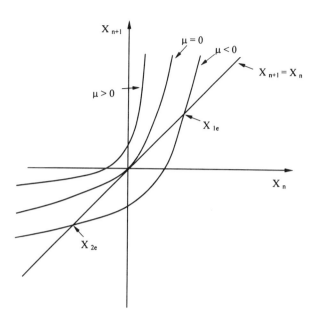

**FIGURE 5.4-2   Mapping function near the bifurcation point
for type I intermittency**

In type III intermittency, the Floquet multiplier is negative and several model equations describing this behavior have been proposed and are discussed in Argyris et al. (1994). In this type of intermittency, unlike in type I, the amplitude of the disturbance increases with time until chaos ensues. Furthermore, the duration of regular motion is longer and theoretically has no upper bound. This type of intermittency has been observed in the Rayleigh-Bénard experiment (Dubois et al., 1983), in an ammonia ring laser (Tang et al., 1991), and in an electronic circuit (Kim et al., 1998). Type II intermittency occurs if a periodic motion becomes unstable due to the crossing of the unit circle by a complex conjugate pair of Floquet multipliers. The imaginary part of the multiplier can be interpreted as a rotation at each iteration or as an additional frequency to the system. Bursts of multi-frequency behavior are observed. Model equations describing this behavior are examined in Argyris et al. (1994). Experimental observations of this type of intermittence are not frequent (Sacher et al., 1989).

Quasi-periodicity

This route to chaos was proposed by Ruelle and Takens (1971) to describe the transition from laminar flow to turbulence. In Chapter 3, we have discussed the Taylor stability problem. We have shown that when the Taylor number exceeds a critical value, the laminar flow breaks down and steady Taylor vortices appear. In Chapter 4, we have introduced a model equation to show that in the non-linear regime, the Taylor vortices become unstable at a second critical Taylor number. We have stated that turbulence sets in after multiple instabilities. This is essentially the route to turbulence as proposed by Landau (reproduced in Hao, 1990) and Hopf (1948). In this scheme, when the laminar flow becomes unstable at a critical Reynolds number Re_1, the bifurcation is of a Hopf type. If Re is only slightly greater than Re_1, the velocity $\underline{v}(\underline{x}, t)$ can be written as

$$\underline{v}(\underline{x}, t) = \underline{v}_0(\underline{x}) + a\,\underline{w}(\underline{x}) \exp[i(\omega_1 t + \alpha_1)] \qquad (5.4\text{-}14)$$

where $\underline{v}_0$ is the velocity of the laminar flow, a is the finite amplitude of the disturbance, ω_1 is the frequency, and α_1 is the initial phase angle.

Equation (5.4-14) is of the form of Equations (3.5-7a-8d) (see also Example 4.2-1). Note also that we have introduced an additional degree of freedom α_1. When $Re \gg Re_1$, Equation (5.4-14) is no longer appropriate and $\underline{v}(\underline{x}, t)$ is represented by a Fourier series and is written as

$$\underline{v}(\underline{x}, t) = \sum_{n=0}^{\infty} a_n\,\underline{f}_n(\underline{x}) \exp[in(\omega_1 t + \alpha_1)] \qquad (5.4\text{-}15)$$

As Re is further increased, $\underline{v}(\underline{x}, t)$ becomes unstable at $Re = Re_2$ and the bifurcation is again of a Hopf type. There are now two frequencies (ω_1, ω_2) and two phases (α_1, α_2) associated with the flow. Equation (5.4-15) is now modified to

$$\underline{v}(\underline{x}, t) = \sum_n \left\{ a_n\,\underline{f}_n \exp[in(\omega_1 t + \alpha_1)] + b_n\,\underline{g}_n \exp[in(\omega_2 t + \alpha_2)] \right\} \qquad (5.4\text{-}16)$$

and the ratio ω_1/ω_2 is an irrational number.

The flow is now a quasi-periodic flow, as discussed earlier (Poincaré section). On further increasing Re, more instabilities occur and at each bifurcation a new independent frequency ω_i and a new phase α_i are introduced. Eventually, after m ($m \longrightarrow \infty$) bifurcations, the flow becomes turbulent and all traces of periodicity have disappeared.

This scenario is now believed to be highly improbable. The number of bifurcations m required to attain turbulence is too large (m is finite). At each bifurcation, we have introduced a perturbation which has to satisfy the equations of motion. This means that we can associate a spatial scale with each perturbation and we can order these spatial scales. At low spatial scales, the magnitudes of the velocity gradients are large which imply that these perturbations decay almost instantaneously through viscous dissipation. These perturbations do not contribute to the establishment of turbulence.

Equation (5.4-16) is a linear superposition of two perturbations. In a non-linear system, there is an interaction between the two perturbations and the original perturbation may not survive. The route proposed by Landau and Hopf does not take into account the interaction between perturbations.

In Landau's model, the motion is essentially quasi-periodic and the mean autocorrelation $\langle a(\tau) \rangle$ [Equation (5.3-26)] does not tend to zero as τ tends to infinity (Table 5.3-1). However, experimental measurements indicate that for turbulent flows, $\langle a(\tau) \rangle$ tends to zero as τ tends to infinity.

The initial stage of the route to turbulence proposed by Ruelle and Takens (1971) is similar to that of Landau. The laminar flow becomes unstable at Re_1 and a periodic flow sets in as described by Equation (5.4-14). This periodic flow (limit cycle) becomes unstable at Re_2 and a new frequency ω_2 and phase β_2 are introduced. The flow is now quasi-periodic on a two-dimensional torus T^2. A further bifurcation leads to a flow on a three-dimensional torus T^3. Newhouse et al. (1978) showed that on T^3, the quasi-periodic motion is unstable and a small disturbance may lead to chaos (turbulence). In this scenario, turbulence can set in after only three bifurcations. Furthermore, the mean autocorrelation $\langle a(\tau) \rangle$ tends to zero as τ tends to infinity. The experimental work of Gollub and Swinney (1975) on Taylor-Couette flow tends to support the model of Ruelle and Takens.

The **circle map** which can be written as

$$\theta_{n+1} = f(\theta_n) = \theta_n + \Omega - (K/2\pi) \sin(2\pi\theta_n) \quad (\text{mod } 1) \qquad (5.4\text{-}17a,b)$$

where Ω and K are constants, is widely used to illustrate the route of quasi-periodicity to chaos.

Here, (mod 1) means we consider only the fractional part of an expression. For example, for the number 1.7 we retain only .7.

The parameter K is a measure of the strength of the non-linearity. If $K = 0$, Equation (5.4-17b) reduces to the linear circle map

$$\theta_{n+1} = \theta_n + \Omega \quad (\text{mod } 1) \qquad (5.4\text{-}18)$$

Equation (5.4-18) can be interpreted as the Poincaré return map of the motion of two oscillators with frequencies ω_1 and ω_2 and $\Omega = \omega_2/\omega_1$. If ω is a rational number, we write

$$\frac{\omega_2}{\omega_1} = \frac{p}{q} \qquad (5.4\text{-}19)$$

where p and q are relative prime integers and we assume that $q \geq p$.

The motion of an oscillator can be decomposed into its Fourier series and for the first oscillator, we can write

$$x_1 = \sum_{n=1}^{\infty} a_n \sin(2\pi n \omega_1 t + \alpha_n) \qquad (5.4\text{-}20)$$

where a_n is the amplitude and α_n is the phase shift.

Similarly, for the second oscillator

$$x_2 = \sum_{n=1}^{\infty} b_n \sin(2\pi n \omega_2 t + \beta_n) \tag{5.4-21}$$

The frequency of the pth harmonic of the first oscillator $(p\omega_1)$ is equal to that of the qth harmonic of the second oscillator $(q\omega_2)$ and this generates a kind of resonance. This phenomenon is the **frequency-locking**. Also after q orbits, the trajectory is closed and the solution is periodic. If Ω is irrational, the motion is quasi-periodic.

Equation (5.4-18) can also be considered to be a movement around the circle in steps of Ω. If Ω is a rational number, after q steps, we return to the original point, that is, a periodic motion. If Ω is an irrational number, we never return to the same point and the motion is quasi-periodic.

The **winding number** w is defined as

$$w = \lim_{n \to \infty} \frac{f^{(n)}(\theta_0) - \theta_0}{n} \tag{5.4-22}$$

where $f^{(n)}$ is the nth iterate and θ_0 is the initial value of θ.

If $K = 0$, combining Equations (5.4-17a, b, 22) yields

$$w = \lim_{n \to \infty} \frac{n\theta_0 + \Omega - \theta_0}{n} = \Omega \tag{5.4-23a,b}$$

The winding number w $(= \Omega)$ is independent of θ_0 and if Ω is a rational number, the solution is periodic. If $K \ne 0$, w depends on the two control parameters Ω and K and for a fixed value of K, there may be a range of values of Ω where w is a rational number. To illustrate this situation, we consider the case $\Omega = 1/2$. If $K = 0$, we have a periodic solution of period 2 and this can be verified from Equation (5.4-18). Note that we are calculating θ_n to modulo 1, that is, $\theta_0 + 1 = \theta_0$. For $K \ne 0$, we determine the relationship between K and Ω such that the solution is of period two. This implies that

$$f^{(2)}(\theta_0) = \theta_0 + 1 \tag{5.4-24}$$

Equations (5.4-17a, b, 24) yields

$$\theta_1 = \theta_0 + \Omega - (K/2\pi)\sin(2\pi\theta_0) \tag{5.4-25a}$$

$$\theta_1 + \Omega - (K/2\pi)\sin(2\pi\theta_1) = \theta_0 + 1 \tag{5.4-25b}$$

Equations (5.4-25a, b) have to be solved numerically. We use a perturbation method to obtain an approximate solution. We assume $K < 1$ and we are interested in the solution near $\Omega = 1/2$. In Equations (5.4-25a, b), we let

$$\Omega = 1/2 + \Delta\Omega \qquad\qquad (5.4\text{-}26)$$

where $\Delta\Omega$ is small.

Substituting Equations (5.4-25a, 26) into Equation (5.4-25b) yields

$$2\Delta\Omega - (K/2\pi)\sin(2\pi\theta_0) - (K/2\pi)\sin[2\pi(\theta_0 + 1/2 + \varepsilon)] = 0 \qquad (5.4\text{-}27a)$$

$$\varepsilon = \Delta\Omega - (K/2\pi)\sin(2\pi\theta_0) \qquad\qquad (5.4\text{-}27b)$$

The quantity ε is small since $\Delta\Omega$ and K are assumed to be small. The term inside the square brackets in Equation (5.4-27a) is expanded in a Taylor series about $2\pi(\theta_0 + 1/2)$ to yield

$$\sin[2\pi(\theta_0 + 1/2 + \varepsilon)] \approx \sin[2\pi(\theta_0 + 1/2)] + 2\pi\varepsilon\cos[2\pi(\theta_0 + 1/2)] + ... \qquad (5.4\text{-}28a)$$

$$\approx -\sin 2\pi\theta_0 - 2\pi[\Delta\Omega - (K/2\pi)\sin(2\pi\theta_0)]\cos 2\pi\theta_0 \qquad (5.4\text{-}28b)$$

Combining Equations (5.4-27a, 28b) yields

$$\Delta\Omega(2 + K\cos 2\pi\theta_0) - (K^2/2\pi)\sin 2\pi\theta_0 \cos 2\pi\theta_0 = 0 \qquad (5.4\text{-}29)$$

If we further assume that $K\Delta\Omega$ is negligible compared to K^2, an approximate solution of Equation (5.4-29) is

$$\Delta\Omega \approx (K^2/4\pi)\sin 2\pi\theta_0 \cos 2\pi\theta_0 \qquad\qquad (5.4\text{-}30a)$$

$$\approx (K^2/8\pi)\sin 4\pi\theta_0 \qquad\qquad (5.4\text{-}30b)$$

Equation (5.4-30b) provides a bound for $\Delta\Omega$. The region in the neighborhood of $\Omega = 1/2$ where a periodic solution exists is approximately given by

$$\Omega = 1/2 \pm K^2/8\pi \qquad\qquad (5.4\text{-}31)$$

This region is the **Arnold tongue** and is shown in Figure 5.4-3.

The stability of the periodic solution is determined by

$$\left|\frac{d}{d\theta} f^{(2)}(\theta_0)\right| < 1 \qquad\qquad (5.4\text{-}32)$$

Equations (5.3-11, 4-17, 32) imply

$$|(1 - K\cos 2\pi\theta_0)(1 - K\cos 2\pi\theta_1)| < 1 \qquad\qquad (5.4\text{-}33)$$

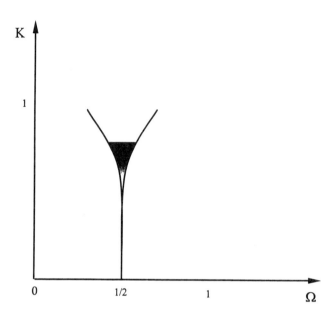

**FIGURE 5.4-3   The Arnold tongue (shaded area) of the circle map
near $\Omega = 1/2$**

If $0 \le K \le 1$ and $\cos 2\pi\theta_i$ (i = 1, 2) are positive, the period solution is stable. We note that the stability of the periodic solution depends on K only. Further, f has an extreme value if

$$f'(\theta) = 1 - K \cos 2\pi\theta = 0 \qquad\qquad (5.4\text{-}34a,b)$$

Equation (5.4-34b) has a solution only if $K > 1$. This implies that if $K > 1$, f is not invertible, that is, one value of θ_{n+1} can be iterated from two possible values of θ_n. In this case, we have chaos. The value of $K = 1$ is a **critical value**.

The circle map has been extensively investigated and a fuller discussion is given in Argyris et al. (1994) and in Problem 1. Here, we summarize the main results.

There is an Arnold tongue associated with every rational value of Ω. The width of the Arnold tongue increases with increasing values of K. If $0 \le K < 1$, the Arnold tongues do not overlap. This implies that there is the possibility of quasi-periodic solutions. For a given value of Ω, the probability of obtaining a quasi-periodic solution decreases with increasing values of K. If $K > 1$, the Arnold tongue overlap and as mentioned earlier the motion is chaotic.

The circle map can describe both periodic and quasi-periodic motions if K is small. On increasing the values of K, a periodic solution may lose its stability and may eventually become chaotic through period-doubling. A quasi-periodic solution becomes chaotic when K crosses the critical line, $K = 1$.

5.5 DISCUSSION

In this chapter, we have described very briefly a few aspects of the theory of chaos, a subject which still attracts a lot of interest. We have seen that chaos can occur in non-linear systems with at least three degrees of freedom. Most of the model equations we have considered have been restricted to sets of three first order ordinary equations. Real systems may have a large number of degrees of freedom. However, Robinson (1998) has shown that the trajectories of any dissipative system can be approximated by the trajectories in an appropriate three-dimensional space. It is therefore not surprising that these model equations adequately describe the chaotic behavior of real systems.

It is still too early to assess the real importance and impact of chaos. As mentioned earlier, chaos has shown that determinism and predictability are not synonymous. Even simple deterministic systems have a horizon of predictability due to the magnification of unavoidable minute errors present in all measurements. We have less confidence in long term forecasting and appreciate the need for diversity.

Equally, the study of chaos has made us realize that determinism and randomness are not mutually exclusive. An apparently random sequence can be generated by a well-defined rule.

Having discovered chaos, it is natural that we attempt to control and make use of it. It is believed that many biological systems operate in the chaotic mode so as to cope better with minor changes. In chemical engineering, mixing in the chaotic regime is more efficient because it allows the fluid elements to explore larger regions. Ott et al. (1994) have suggested that to keep a system in chaotic motion and to avoid the windows of periodic motion we need to apply piecewise smooth changes and not continuous changes.

The theory of chaos has brought an improvement to modeling and has led to a possible description of complex systems. However, it does not always provide all the information we require. For example, the route to turbulence described by Ruelle and Takens (1971) might be more realistic than that of Landau (see Hao, 1990) but it does not allow us to calculate a drag force or a flow rate, quantities which are of interest to engineers.

PROBLEMS

1. Compute the values of θ_n as given by the circle map [Equations (5.4-17a, b)] for $\Omega = 0.53$, $K/2\pi = 0.62$, and for two initial values $\Omega_0 = 0.06$ and $\Omega_0 = 0.07$.

    Verify that starting with $\Omega_0 = 0.06$, we have a cycle of period 2 whereas with $\Omega_0 = 0.07$, we have a cycle of period 4. Calculate θ_n for the same values of Ω and K and for several initial values lying between 0.06 and 0.07. What conclusions can you draw?

2. Determine the equilibrium points of the quadratic map

$$x_{n+1} = \mu - x_n^2$$

Compute x_n for the following values of μ: 0.5, 0.8, 1.5, and 2. Choose $x_0 = 0.1$. For $\mu = 1.5$, choose another value of x_0 which is close to 0.1. What conclusions can you draw? For what values of μ does a periodic solution exist?

3. Rössler's (1976) equations can be written as

$$\dot{x}_1 = -(x_2 + x_3)$$

$$\dot{x}_2 = x_1 + a x_2$$

$$\dot{x}_3 = b + x_3 (x_1 - c)$$

Examine the properties of the solutions for $a = b = 0.2$ and for various values of c. Note that $x_1 x_3$ is the only non-linear term. Consider first the two-dimensional case ($x_3 = 0$); next consider the three-dimensional problem.

4. Compute a Liapunov exponent from the data given in Table 5.2-1. This can be done in the following manner. From the sequence x_n, select x_i and x_j such that their values are close ($x_i \approx x_j$). Denote $s_0 = |x_j - x_i|$, $s_1 = |x_{j+1} - x_{i+1}|$, ... , $s_m = |x_{j+m} - x_{i+m}|$. If the trajectories diverge exponentially, s_m can be written as

$$s_m = s_0 e^{m\lambda}$$

On taking the ℓn, we obtain

$$\lambda = (1/m) \ell n (s_m/s_0)$$

Plot ℓn s_m versus m. If it is a straight line, the slope is a Liapunov exponent.

Evaluate λ from Equation (5.3-13). Establish whether there is an agreement between the two methods.

5. The construction of a Koch curve is similar to that of the Cantor set. We start by dividing a straight line of unit length into three equal parts. The middle part is removed and is replaced by two segments each of length 1/3 so as to form a triangular tent in the middle. Now we have 4 segments, each of length 1/3. We repeat the same process on each segment, that is, divide each segment into three equal parts, replace the middle part by a triangular tent, and this results in a figure of 16 segments, each of length 1/9. After n operations, we obtain a figure of 4^n segments, each of length $(1/3)^n$. Calculate the box dimension D_c of the Koch curve.

6.   The Schwarzian derivative $SD[f(x)]$ is defined by

$$SD[f(x)] = f'''/f' - (3/2)(f''/f')^2$$

where f', f'', and f''' are the first, second, and third derivatives respectively.

Calculate the $SD[f(x)]$ of the logistic map [Equation (5.2-1)] and of the circle map [Equation (5.4-17)]. Note that $SD[f(x)]$ is negative in the interval $[0, 1]$ for both maps and this is a necessary condition for the occurrence of period doubling.

7.   The forward iteration of the Hénon map is given by Equations (5.3-42a, b). Show that the backward iteration is given by

$$x_n = y_{n+1}/b$$

$$y_n = x_{n+1} - 1 + (a/b^2)y_{n+1}^2$$

Note that the Hénon map is invertible, that is, for a given (x_n, y_n), we have a unique (x_{n-1}, y_{n-1}). Is this true for the logistic map [Equation (5.2-1)]?

Compute a few forward and backward iterations for $b = 0.3$ and $a = 1.4$. Choose more than one initial value. Do the backward orbits escape to infinity?

8.   Graham (1995) has proposed a model to describe the complicated non-periodic flow of polymer melts between two oscillating parallel plates. In this model, the no-slip boundary condition is assumed to be invalid and the slip velocity u_s is given by

$$u_s + \lambda_s \dot{u}_s = \phi(\tau)$$

where τ is the shear stress and λ_s is a constant.

Two viscoelastic models are considered, namely the Maxwell and the White-Metzner models. The relevant equations are solved numerically. From the results presented by the author, deduce whether the flow is quasi-periodic or chaotic?

REFERENCES

ALEKSIC, Z., Physica, **52D**, 362 (1991).

ANDERECK, C.D., LIU, S., and SWINNEY, H.L., J. Fluid Mech., **164**, 155 (1986).

ARGOUL, F., ARNEODO, A., RICHETTI, P., and ROUX, J.C., Accounts of Chem. Res., **20**, 436 (1987).

ARGYRIS, J., FAUST, G., and HAASE, M., An Exploration of Chaos (North Holland, Amsterdam, 1994).

ARNOLD, V.I., Geometrical Methods in the Theory of Ordinary Differential Equations (Springer-Verlag, Berlin, 1983).

BARNETT, S., Matrices (Clarendon Press, Oxford, 1990).

BENJAMIN, T.B., Proc. Roy. Soc. Lond., **A359**, 1 (1978).

BERGÉ, P., DUBOIS, M., MANNEVILLE, P., and POMEAU, Y., J. de Phys. Lett., **41**, 341 (1980).

BERGER, J.S. and PERLMULTER, D.D., A.I.Ch.E. J., **10**, 233 (1964).

BIRD, R.B., STEWART, W.E., and LIGHTFOOT, E.N., Transport Phenomena (Wiley, New York, 1960).

CAMERON, A., Basic Lubrification Theory (Longman, London, 1971).

CASTI, J., Ecological Modelling, **14**, 293 (1982).

CESARI, L., Asymptotic Behavior and Stability Problems in Ordinary Differential Equations (Springer-Verlag, Berlin, 1958).

CHAN MAN FONG, C.F., KALONI, P.N., and DE KEE, D., Acta Mech., **115**, 231 (1996).

CHAN MAN FONG, C.F., DE KEE, D., and KALONI, P.N., Advanced Mathematics for Applied and Pure Sciences (Gordon and Breach, Amsterdam, 1997).

CHANDRASEKHAR, S., Hydrodynamic and Hydromagnetic Stability (Oxford University Press, Oxford, 1961).

CHURCHILL, R.C., PECELLI, G., and ROD, D.L., Lecture Notes in Physics, Vol. 93, edited by G. Casati and J. Ford (Springer-Verlag, Berlin, 1979).

CLOSE, C.M. and FREDERICK, D.K., Modeling and Analysis of Dynamical Systems (Houghton Mifflin, Boston, 1978).

COFFMAN, K.G., MCCORMICK, W.D., NOSZTICZIUS, Z., SIMOYI, R.H., and SWINNEY, H.L., J. Chem. Phys., **86**, 119 (1987).

COLES, D., J. Fluid Mech., **21**, 385 (1965).

CRANK, J., The Mathematics of Diffusion (Clarendon Press, Oxford, 1975).

DAVEY, A., J. Fluid Mech., **14**, 336 (1962).

DIACU, F. and HOLMES, P., Celestial Encounters. The Origins of Chaos and Stability (Princeton University Press, Princeton, New Jersey, 1996).

DI PRIMA, R.C., J. Lub. Tech., **90**, 173 (1968); **91**, 45 (1969).

DRAZIN, P.G. and REID, W.H., Hydrodynamic Stability (Cambridge University Press, Cambridge, 1981).

DUBOIS, M. and BERGÉ, P., J. de Physique, **42**, 167 (1981).

DUBOIS, M., RUBIO, M.A., and BERGÉ, P., Phys. Rev. Lett., **51**, 1446, 2345 (1983).

FARMER, J.D., OTT, E., and YORKE, J.A., Physica, **7D**, 153 (1983).

FARMER, J.D., CRUTCHFIELD, J., FROEHLING, H., PACKARD, N., and SHAW, R., Annals New York Academy of Sciences, **357**, 453 (1980).

FEIGENBAUM, M.J., J. Stat. Phys., **19**, 160 (1978).

GIBBS, H.M., HOPF, F.A., KAPLAN, D.L., and SHOEMAKER, R.L., Phys. Rev. Lett., **46**, 474 (1981).

GILMORE, R., Phys. Rev., **A20**, 2510 (1979).

GLEICK, J., Chaos, Making a New Science (Viking, New York, 1987).

GLENDINNING, P., Stability, Instability, and Chaos (Cambridge University Press, Cambridge, 1994).

GOLDSTEIN, H., Classical Mechanics, 2nd Ed. (Addison-Wesley, Reading, 1980).

GOLLUB, J.P. and SWINNEY, H.L., Phys. Rev. Lett., **35**, 927 (1975).

GRAHAM, M.D., J. Rheol., **39**, 697 (1995).

GRASSHERGER, P. and PROCACCIA, I., Phys. Rev., **A28**, 2591 (1983).

GUREL, O. and LAPIDUS, L., I. & E.C. Chem., **61**, 30 (1969).

HAHN, W., Theory and Application of Liapunov's Direct Method (Prentice Hall, New Jersey, 1963).

HALE, J.K. and KOÇAK, H., Dynamics and Bifurcations (Springer-Verlag, Berlin, 1991).

HANKS, T.C., J. Geophys. Res., **76**, 537 (1971).

HAO, B.L. (Ed.), <u>Chaos</u> (World Scientific, Singapore, 1990).

HÉNON, M., <u>Commun. Math. Phys.</u>, **50**, 69 (1976).

HÉNON, M., <u>Physica</u>, **5D**, 412 (1982).

HÉNON, M. and HEILES, C., <u>Astrophys. J.</u>, **69**, 73 (1964).

HILBORN, R.C., <u>Chaos and Nonlinear Dynamics</u> (Oxford University Press, Oxford, 1994).

HOLMES, P.J. and RAND, D.A., <u>J. Sound and Vibrations</u>, **44**, 237 (1976).

HOPF, E., <u>Commun. on Pure and Appl. Math.</u>, **1**, 303 (1948).

HSIEH, D.Y. and HO, S.P., <u>Waves and Stability in Fluids</u> (World Scientific, Singapore, 1994).

HSU, J.C. and MEYER, A.U., <u>Modern Control Principles and Applications</u> (McGraw-Hill, New York, 1968).

JORDAN, D.W. and SMITH, P., <u>Nonlinear Ordinary Differential Equations</u>, 2nd Ed. (Clarendon Press, Oxford, 1987).

KADANOFF, L., <u>Physics Today</u>, **36**, 46 (December 1983).

KENNEL, M.B., BROWN, R., and ABARBANEL, H.D.I, <u>Phys. Rev.</u>, **A56**, 3403 (1992).

KEVORKIAN, J. and COLE, J.D., <u>Perturbation Methods in Applied Mathematics</u> (Springer-Verlag, Berlin, 1981).

KIM, C.M., YIM, G.S., RYU, J.W., and PARK, Y.J., <u>Phys. Rev. Lett.</u>, **80**, 5317 (1998).

KRYLOV, N. and BOGOLIUBOV, N.N., <u>Introduction to Nonlinear Mechanics</u> (Princeton University Press, Princeton, New Jersey, 1947).

LARSON, R.E. and EDWARDS, B.H., <u>Elementary Linear Algebra</u>, 2nd Ed. (D.C. Heath, Toronto, 1991).

LASALLE, J. and LEFSCHETZ, S., <u>Stability by Liapunov's Direct Method</u> (Academic Press, New York, 1961).

LEFEVER, R., NICOLIS, G., and BORCKMANS, P., <u>J. Chem. Soc., Faraday Trans.</u>, **84**, 1013 (1988).

LETOV, A.M., <u>Stability in Nonlinear Control Systems</u> (Princeton University Press, Princeton, New Jersey, 1961).

LIBCHABER, A., FAUVE, S., and LAROCHE, C., <u>Physica</u>, **7D**, 73 (1983).

LIEBERT, W. and SCHUSTER, H.G., <u>Phys. Lett.</u>, **A142**, 107 (1989).

LIGHTHILL, J., <u>Proc. Roy. Soc. Lond.</u>, **A407**, 35 (1986).

LORENZ, E.N., <u>J. Atmos. Sci.</u>, **20**, 130 (1963).

LOTKA, A.J., <u>Elements of Mathematical Biology</u> (Dover, New York, 1956).

MANDELBROT, B.B., The Fractal Geometry of Nature (W.H. Freeman, New York, 1982).

MAY, R.M., Nature, **261**, 459 (June 1976).

MICHAELIS, L. and MENTEN, M.I., Biochem. Z., **49**, 333 (1913).

MINORSKY, N., Nonlinear Oscillations (Van Nostrand, New York, 1962).

MORRISON, J.A., SIAM Rev., **8**, 66 (1966).

MURRAY, J.D., Mathematical Biology, 2nd Ed. (Springer-Verlag, Berlin, 1993).

NAYFEH, A.H., Perturbation Methods (Wiley, New York, 1973).

NEWHOUSE, S., RUELLE, D., and TAKENS, F., Commun. Math. Phys., **64**, 35 (1978).

ODUM, E.P., Fundamentals of Ecology, 3rd Ed. (W.B. Saunders, Philadelphia, 1971).

OTT, E., SAUER, T., and YORKE, J.A., Coping with Chaos: Analysis of Chaotic Data and the Exploitation of Chaotic Systems (Wiley, New York, 1994).

POINCARÉ, H., Les méthodes nouvelles de la mécanique céleste (Gauthier-Villars, Paris, 1892).

POLLARD, H., Mathematical Introduction to Celestial Mechanics (Prentice-Hall, New Jersey, 1966).

POMEAU, Y., ROUX, R.C., ROSSI, A., BACHELART, S., and VIDAL, C., J. de Phys. Lett., **42**, 271 (1981).

PRIGOGINE, I. and LEFEVER, R., J. Chem. Phys., **48**, 1695 (1968).

PUJAR, N.S. and ZYDNEY, A.L., A.I.Ch.E. J., **42**, 2101 (1996).

REYNOLDS, O., Phil. Trans. Roy. Soc., **174**, 935 (1883).

ROBERTS, D.V., Enzyme Kinetics (Cambridge University Press, Cambridge, 1977).

ROBINSON, J.C., Nonlinearity, **11**, 529 (1998).

ROSENHEAD, L. (Ed.), Laminar Boundary Layer (Clarendon Press, Oxford, 1963).

RÖSSLER, O.E., Phys. Lett., **A71**, 155 (1976).

ROUCHE, N., HABETS, P., and LALOY, M., Stability Theory by Liapunov's Direct Method (Springer-Verlag, Berlin, 1977.)

ROUX, J.C., SIMOYI, R.H., and SWINNEY, H.L., Physica, **8D**, 257 (1983).

RUELLE, D. and TAKENS, F., Commun. Math. Phys., **20**, 167 (1971).

SACHER, J., ELSÄSSER, W., and GÖBEL, E., Phys. Rev. Lett., **63**, 2224 (1989).

SANDEFUR, J.T., Discrete Dynamical Systems (Clarendon Press, Oxford, 1990).

SAUNDERS, P.T., An Introduction to Catastrophe Theory (Cambridge University Press, Cambridge, 1980).

SPARROW, C., The Lorenz Equations (Springer-Verlag, Berlin, 1982).

STEWART, I., <u>Does God Play Dice? The Mathematics of Chaos</u> (Blackwell, New York, 1989).

SUSSMAN, H.J. and ZAHLER, R.S., <u>Synthèse</u>, **37**, 117 (1978).

SWINNEY, H.L. and GOLLUP, J.P., <u>Hydrodynamic Instabilities and the Transition to Turbulence, Vol. 45 - Topic in Appl. Phys.</u> (Springer-Verlag, Berlin, 1981).

TAKENS, F., <u>Dynamical Systems and Turbulence, Vol. 898 - Lecture Notes in Mathematics</u> (Springer-Verlag, Berlin, 1981).

TANG, D.Y., PUJOL, J., and WEISS, C.O., <u>Phys. Rev.</u>, **A44**, 35 (1991).

TAYLOR, G.I., <u>Phil. Trans. Roy. Soc.</u>, **A223**, 289 (1923).

TESTA. J., PEREZ, J., and JEFFRIES, C., <u>Phys. Rev. Lett.</u>, **48**, 714 (1982).

THOM, R., <u>Stabilité structurelle et morphogénèse</u> (W.A. Benjamin, Reading, 1972).

THOM, R., <u>Structural Stability, the Theory of Catastrophes and Applications in the Sciences, Vol. 525 - Lecture Notes in Mathematics</u>, edited by P. Hilton (Springer-Verlag, Berlin, 1976).

THOMPSON, J.M.T., <u>Instabilities and Catastrophes in Science and Engineering</u> (Wiley, New York, 1982).

VAN DYKE, M., <u>Perturbation Methods in Fluid Mechanics</u> (Parabolic Press, Stanford, California, 1975).

VAN DYKE, M., <u>Adv. Appl. Mech.</u>, **25**, 1 (1987).

VARIAN, H.R., <u>Economic Inquiry</u>, **17**, 14 (1979).

WILCOX, D.C., <u>Perturbation Methods in the Computer Age</u> (DCW Industries Inc., La Canada, California, 1995).

WOLF, A., SWIFT, J.B., SWINNEY, H.L., and VASTANO, J.A., <u>Physica</u>, **16D**, 285 (1985).

WOLLKIND, D.J., <u>SIAM Rev.</u>, **19**, 502 (1977).

WOODCOCK, A.E.R. and POSTON, T., <u>A Geometrical Study of the Elementary Catastrophes, Vol. 373 - Lecture Notes in Mathematics</u> (Springer-Verlag, Berlin, 1974).

WU, X. and XIONG, K., <u>Int. J. Control</u>, **69**, 353 (1998).

ZAHLER, R.S. and SUSSMAN, H.J., <u>Nature</u>, **269**, 759 (October 1977).

ZEEMAN, E.C., <u>Scientific American</u>, **234**, 65 (April 1976).

ZEEMAN, E.C., <u>Catastrophe Theory: Selected Papers 1972-1977</u> (Addison-Wesley, Reading, 1977).

AUTHOR INDEX

SUBJECT INDEX